Partial Differential Equations of First Order and Their Applications to Physics

Second Edition

Partial Differential Equations of First Order and Their Applications to Physics

Second Edition

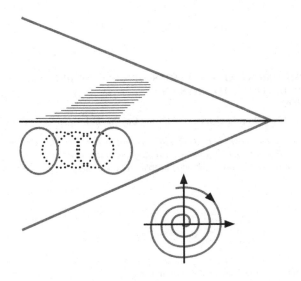

Gustavo López Velázquez

University of Guadalajara, Mexico

World Scientific

NEW JERSEY · LONDON · SINGAPORE · BEIJING · SHANGHAI · HONG KONG · TAIPEI · CHENNAI

Published by

World Scientific Publishing Co. Pte. Ltd.

5 Toh Tuck Link, Singapore 596224

USA office: 27 Warren Street, Suite 401-402, Hackensack, NJ 07601

UK office: 57 Shelton Street, Covent Garden, London WC2H 9HE

British Library Cataloguing-in-Publication Data
A catalogue record for this book is available from the British Library.

PARTIAL DIFFERENTIAL EQUATIONS OF FIRST ORDER AND THEIR APPLICATIONS TO PHYSICS
Second Edition

ISBN-13 978-981-4390-37-8
ISBN-10 981-4390-37-2

Printed in Singapore by World Scientific Printers.

I want to dedicate this book to my parents, *J. Guadalupe López López and Vicenta Velázquez*, for their support and their love during their lifetime.

I also want to dedicate this book to my sister *Esperanza* and my friend *Javier I. Hernández* who both passed away and their life were an inspiration to me.

I finally want to dedicate this book to my granddaughter *Jezen* for being a wonderful and valuable gift to our life.

Preface

This book is a collection of notes of the course *Partial Differential Equations of First Order* given by *Dr. Gustavo López Velázquez* during different periods of time at different universities. The main body of this book was put together during a seminar, given by Dr. G. López in collaboration with M.A. Murguía and M. Romero at the Physics Institute of the University of Guanajuato, from August to November of 1988. This collection was made with the help of M.A. Murguía, M. Romero, E. Benitez, and C. Melo. The final version of the first edition was made by Dr. López at the University of Guadalajara, México, and at Los Alamos National Laboratory (Department of Non Linear Dynamics), New Mexico, USA, from 1995 to 1996. I want to point out that without the help and enthusiasm of M.A. Murguía, M. Romero and C. Melo, the elaboration of these notes would not have been possible. In addition, I want to thank also to Alejandra Taylor for her collaboration during the revision of the text.

For the second edition, I want to thank to Gustavo Montes for his help for reviewing and typing much of the material of this new version. I expanded the contain of each chapter adding new material which I considered relevant for the students, I did a lot of typing of corrections that appeared in the first edition, for which I apologize to the readers. Finally, I have included several exercises in each chapter which can help readers to master each topic.

In this book I try to point out the mathematical importance of Partial Differential Equations of First Order in Physics and Applied Sciences. The intention of this book is to give to mathematicians a wide view of the applications of this branch in Physics, and to give to physicists and applied scientists a powerful tool for solving some problems appearing in Classical Mechanics, Quantum Physics, Optics, and General Relativity. This book is intended for senior or first year graduate students in mathematics, physics or engineering curricula.

Mathematics is a gift...
for man to understand the laws of nature
which make up the whole Universe.

G. López

Contents

Chapter 1

Geometric Concepts and Generalities

In this chapter we shall study some geometric concepts that are basic to understand the geometric meaning of the partial differential equation in $\mathbb{R}^3$.

1.1 Surfaces and Curves in Three Dimensions

By a surface S in $\mathbb{R}^3$ we mean any relation between the rectangular cartesian coordinates (x, y, z) of a point in this space given by following expressions

$$\text{(explicit)} \quad z = f(x, y) \tag{1.1}$$

$$\text{(implicit)} \quad F(x, y, z) = 0 \tag{1.2}$$

$$\text{(parametric)} \quad x = f_1(u, v), \quad y = f_2(u, v), \quad z = f_3(u, v) \tag{1.3}$$

where to each pair of values of u, v there corresponds a set of numbers (x, y, z) and hence a point in space. While the expressions for the surfaces (1.1) and (1.2) are unique, the parametric expression (1.3) is not unique. For example, the spherical surface $x^2 + y^2 + z^2 = a^2$, (see Fig. 1.1) can be parameterized by

$$x = a \sin u \cos v, \quad y = a \sin u \sin v, \quad z = a \cos u,$$

or

$$x = a \frac{1 - v^2}{1 + v^2} \cos u, \quad y = a \frac{1 - v^2}{1 + v^2} \sin u, \quad z = \frac{2ab}{1 + v^2},$$

or more general

$$x = a \frac{1 - g(v)}{1 + g(v)} \cos u, \quad y = a \frac{1 - g(v)}{1 + g(v)} \sin u, \quad z = \frac{2a\sqrt{g(v)}}{1 + g(v)},$$

1

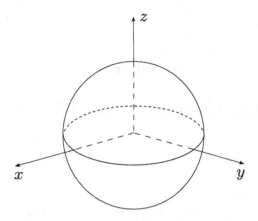

Fig. 1.1   Surface of sphere is not unique parameterized.

where $g(v)$ is such that $1 + g(v) > 0$ for all $v \in \mathbb{R}$. By a curve Γ in $\mathbb{R}^3$ we understand any relation between a point (x, y, z) in this space of the form

$$\text{(non-parametric)} \quad f(x, y, z) = 0; \quad g(x, y, z) = 0 \qquad (1.4)$$

$$\text{(parametric)} \quad x = f_1(t), \quad y = f_2(t), \quad z = f_3(t) \qquad (1.5)$$

where t is a continuous variable called the parameter of the curve (a usual parameter is the length of a curve measured from some fixed point). The relation (1.4) expresses in fact the intersection of two surfaces (see Fig. 1.2).

A surface can be thought of as being generated by a set of curves in the following way, the surface $f(x, y, z) = 0$ is generated by set of curves Γ_k defined by

$$x = k, \quad f(x, y, k) = 0 \qquad (1.6)$$

where k takes a certain interval of values, for example, the sphere (see Fig. 1.3) can be seen as generated by the curves Γ_k given by

$$z = k, \quad x^2 + y^2 = a^2 - k^2 \quad \text{with} \quad -a \le k \le a .$$

Let a curve Γ be parameterized by the length of the curve s, an let

$$P_0 = (x(0), y(0), z(0)),$$

$$P = (x(s), y(s), z(s)),$$

and

$$Q = (x(s + \delta s), y(s + \delta s), z(s + \delta s))$$

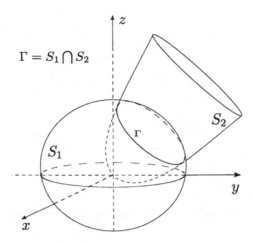

Fig. 1.2   The intersection of the surfaces S_1 and S_2 generated the curve Γ.

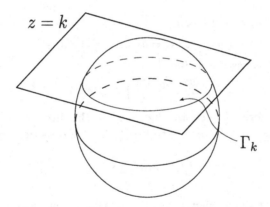

Fig. 1.3   Surface of sphere is generated by the planes $z = k$, $-a \leq k \leq a$.

be three points on Γ (see Fig. 1.4). If δc is the Euclidean distance between the point P and Q, we restrict ourselves to those kind of curves which satisfy

$$\lim_{\delta s \to 0} \frac{\delta c}{\delta s} = 1. \qquad (1.7)$$

This means, for example that we will not be interested in such a curves turn around and cross themselves in some point. The directions cosines of

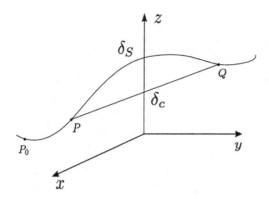

Fig. 1.4   Distances δc and δs between the points P and Q.

the chord PQ are

$$\left(\frac{x(s + \delta s) - x(s)}{\delta c}, \frac{y(s + \delta s) - y(s)}{\delta c}, \frac{z(s + \delta s) - z(s)}{\delta c} \right),$$

and due to the Taylor's theorem,

$$x(s + \delta s) - x(s) = \delta s(dx/ds) + O(\delta s^2),$$

these direction cosines are deduced to

$$\frac{\delta s}{\delta c} \left(\frac{dx}{ds}, \frac{dy}{ds}, \frac{dz}{ds} \right) + O(\delta s^2),$$

and δs tends to zero. The chord PQ takes up the direction of the tangent to the curve at P, and according to Eq. (1.7), the direction cosines of this tangent are

$$\left(\frac{dx}{ds}, \frac{dy}{ds}, \frac{dz}{ds} \right). \tag{1.8}$$

By a curve Γ given in parametric form, with s as the parameter, and passing upon a surface S given by expression (1.2) (see Fig. 1.5), we understand that the following identity is satisfied

$$F(x(s), y(s), z(s)) = 0 \tag{1.9}$$

for all the values s in the curve which lies on the surface.

If Eq. (1.9) is satisfied for all values of s, then the curve lies completely on the surface. Of course, if the curve is caused by the intersection of two surfaces, this curve lies completely on both surfaces. Differentiating Eq. (1.9) with respect to s, we obtain

$$\frac{\partial F}{\partial x} \frac{dx}{ds} + \frac{\partial F}{\partial y} \frac{dy}{ds} + \frac{\partial F}{\partial z} \frac{dz}{ds} = 0. \tag{1.10}$$

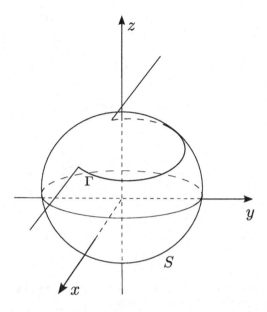

Fig. 1.5   Part of a curve Γ laying on the sphere.

From the relation (1.8) and (1.10) we see that the tangent to the curve Γ at any point P on the surface S is perpendicular to the gradient of F

$$\nabla F = \left(\frac{\partial F}{\partial x}, \frac{\partial F}{\partial y}, \frac{\partial F}{\partial z}\right), \tag{1.11}$$

and this is true for any curve Γ lying on S passing through P, then the vector ∇F is normal to the surface S at the point P (see Fig. 1.6), and the normal vector at any point of the surface is given by this gradient, valued at this point, divided by its magnitude. If the equation of the surface S is given in the form $z = f(x, y)$, defining p and q as

$$p = \frac{\partial z}{\partial x}, \quad \text{and} \quad q = \frac{\partial z}{\partial y}, \tag{1.12}$$

and making $F = f(x, y) - z$, it follows that $F_x = p$, $F_y = q$, $F_z = -1$ and the unitary vector $\hat{n}$ normal to the surface in any point is

$$\hat{n} = \frac{1}{[p^2 + q^2 + 1]^{1/2}} (p, q, -1). \tag{1.13}$$

Let $P = (x, y, z)$ be a point on the surface S defined by $F(x, y, z) = 0$ and let π_1 be the tangent plane at this point, if (X, Y, Z) is any other point on π_1 then, from the above discussion, the vector $(X - x, Y - y, Z - z)$ lying

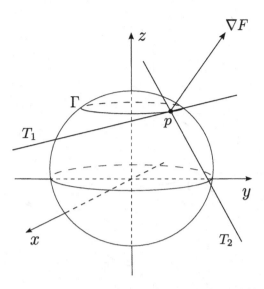

Fig. 1.6   Normal vector ∇F at the point P of the sphere, and the tangent T_1 and orthogonal T_2 lines of to the curve Γ at this point.

on the plane π_1, must be perpendicular to the normal directions ∇F at P, so the equation tangent plane π_1, (see Fig. 1.7) is

$$(X - x)\frac{\partial F}{\partial x} + (Y - y)\frac{\partial F}{\partial y} + (Z - z)\frac{\partial F}{\partial z} = 0. \qquad (1.14)$$

Similarly, let S_2 be other surface defined by $G(x, y, z) = 0$ which intersects the surface S_1 generating a curve Γ that passes through the point P. The equation for the tangent plane π_2 of the surface at the point P is

$$(X' - x)\frac{\partial G}{\partial x} + (Y' - y)\frac{\partial G}{\partial y} + (Z' - z)\frac{\partial G}{\partial z} = 0, \qquad (1.15)$$

where (X', Y', Z') is now any other point on this tangent plane π_2 (see Fig. 1.8). The equation of the line L generated by the intersection of both planes π_1 and π_2 must be such that its directions cosines vector $(X'' - x, Y'' - y, Z'' - z)$, where (X'', Y'', Z'') is now any other point on the line L, is perpendicular to ∇F and ∇G, that is, it must be parallel to the cross product of ∇F with ∇G,

$$\nabla F \times \nabla G = \left(\frac{\partial F}{\partial y}\frac{\partial G}{\partial x} - \frac{\partial F}{\partial z}\frac{\partial G}{\partial y}, \frac{\partial F}{\partial z}\frac{\partial G}{\partial x} - \frac{\partial F}{\partial x}\frac{\partial G}{\partial z}, \frac{\partial F}{\partial x}\frac{\partial G}{\partial y} - \frac{\partial F}{\partial y}\frac{\partial G}{\partial x} \right),$$
$$(1.16)$$

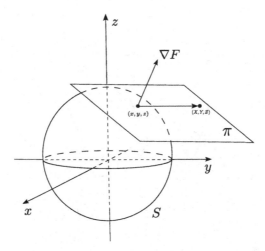

Fig. 1.7 Tangent plane π of the surfaces S and S_2 at point $P = (x, y, z)$, and vector joining this point with other point, (X, Y, Z), on the plane.

and therefore is proportional to this vector, establishing the following equations

$$\frac{X'' - x}{\dfrac{\partial(F, G)}{\partial(y, z)}} = \frac{Y'' - y}{\dfrac{\partial(F, G)}{\partial(z, x)}} = \frac{Z'' - z}{\dfrac{\partial(F, G)}{\partial(x, y)}}, \qquad (1.17)$$

where $\partial(FG)/\partial(y, z)$ is given by

$$\frac{\partial(F, G)}{\partial(y, z)} = det \begin{pmatrix} F_y & F_z \\ G_y & G_z \end{pmatrix} = F_y G_z - F_z G_y, \qquad (1.18)$$

and so on. Choosing the point on L close enough to (x, y, z), i.e. $X'' = x + dx, Y'' = y + dy, Z'' = z + dz$, and given F and G, then (1.17) has the following form

$$\frac{dx}{P(x, y, z)} = \frac{dy}{Q(x, y, z)} = \frac{dz}{R(x, y, z)}, \qquad (1.19)$$

where P, Q and R are known functions. The solution of (1.19) gives us the lines with tangent parallel to the vector field $(P(x, y, z), Q(x, y, z), R(x, y, z))$. These integral curves form a two-parameter family of curves in three dimensional space.

Example 1.1. Give the tangent planes at the point $P = (0, (\sqrt{17}/2 - 1/2)^{1/2}, (\sqrt{17} - 1)/2)$ of the surface $x^2 + y^2 + z^2 = 4$. Give their normal vectors and the equation of the tangent line generated by the intersection

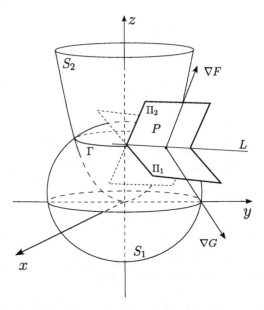

Fig. 1.8   Interception of the tangent planes π_1 and π_2 to the surfaces S_1 and S_2 at the point P, generating the tangent line L.

of the planes at this point. Give the curve Γ generated by the intersection of both surfaces.

In this case

$$F = x^2 + y^2 + z^2 - 4 = 0$$

and

$$G = x^2 + y^2 = 0.$$

$\nabla F = (2x, 2y, 2z)$, the vector normal of the surface F at P is

$$\hat{n}_1 = \frac{1}{4}[0, (2\sqrt{17} - 2)^{1/2}, \sqrt{17} - 1],$$

$\nabla G = (2x, 2y, -1)$, the normal vector of the surface G at P is

$$\hat{n}_2 = \frac{1}{(2\sqrt{17} - 1)^{1/2}}[0, (2\sqrt{17} - 2)^{1/2}, -1].$$

$\dfrac{\partial(F,G)}{\partial(y,z)} = -2y(1+2z)$, $\dfrac{\partial(F,G)}{\partial(z,x)} = 2x(1+2z)$, and $\dfrac{\partial(F,G)}{\partial(x,y)} = 0$. Therefore, at the point P they have the values $-[34(\sqrt{17} - 1)]^{1/2}, 0$ and 0 respectively.

The equations of the planes at the point P are according to Eq. (1.14) and Eq. (1.15)

$$(2\sqrt{17} - 2)^{1/2}Y + (\sqrt{17} - 1)Z = 8,$$

where Y, Z are coordinates of the plane π_1 and

$$(2\sqrt{17} - 2)^{1/2}Y' - Z' = \frac{\sqrt{17} - 1}{2},$$

where Y', Z' are coordinates of the plane π_2. The equations for the line lying on both tangent planes which is tangent to the surface at the point P is given, according to (1.17), by the equations

$$\frac{X''}{-[34(\sqrt{17} - 1)]^{1/2}} = \frac{Y'' - [(\sqrt{17} - 1)/2]^{1/2}}{0} = \frac{Z'' - [(\sqrt{17} - 1)/2]^{1/2}}{0},$$

or writing these equations in parametric way, it follows

$$X'' = -[34(\sqrt{17}-1)]^{1/2}s, \quad Y'' = [(\sqrt{17}-1)/2]^{1/2}, \quad Z'' = [(\sqrt{17}-1)/2]^{1/2}$$

(see Fig. 1.9). The equation of the curve Γ, which is generated by the intersection both surfaces and passes for the point P, is given by

$$x^2 + y^2 = \frac{\sqrt{17} - 1}{2}.$$

1.2 Parallelism of Vector Fields

One of the important mathematical notion for the rest of the book is that one of parallelism of vector fields.

Definition 1.1. Let $\mathbf{a}, \mathbf{b}$ be two vector fields continuously defined in region $\Omega \subset \mathbb{R}^3$, one says that these vector fields are parallel at $P \in \Omega$, $P = (x, y, z)$, if and only if there is an scalar continuous function λ defined also in Ω such that

$$\mathbf{a}(P) = \lambda(P)\mathbf{b}(P) , \tag{1.20}$$

and one will say that these vector fields are parallel on Ω if for any $P \in \Omega$, this relation is satisfied. In the case of $\mathbf{a}$ and $\mathbf{b}$ being constant, λ is just a real number.

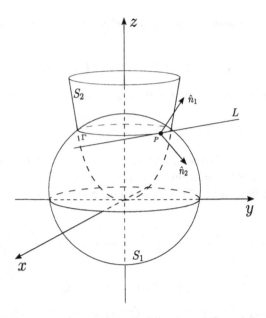

Fig. 1.9   Normal vectors $\hat{n}_1$ and $\hat{n}_2$ to the surfaces S_1 and S_2 and the tangent line L at the point P of their interception.

Assume that $\mathbf{a} = (a_1, a_2, a_3)$ and $\mathbf{b} = (b_1, b_2, b_3)$, then, the above relation can be written as

$$a_1 = \lambda b_1, \tag{1.21a}$$

$$a_2 = \lambda b_2, \tag{1.21b}$$

$$a_3 = \lambda b_3, \tag{1.21c}$$

or in the following way

$$\frac{a_1}{b_1} = \frac{a_2}{b_2} = \frac{a_3}{b_3}. \tag{1.22}$$

This last expression is very common and is good to keep it in mind.

Observation 1.1. Multiplying expression (1.21a) by an arbitrary continuous function $\mu(x, y, z)$, the expression (1.21b) by another arbitrary continuos function $\nu(x, y, z)$ and add them, we would get

$$\mu a_1 + \nu a_2 = \lambda(\mu b_1 + \nu b_2) \tag{1.23a}$$

$$a_3 = \lambda b_3 , \tag{1.23b}$$

for any $\mathbf{a}$ parallel to $\mathbf{b}$ in Ω, which can be written as

$$\frac{\mu a_1 + \nu a_2}{\mu b_1 + \nu b_2} = \frac{a_3}{b_3}. \tag{1.24}$$

Observation 1.2. Assume that one of the components of the vector filed **b** is zero, says b_3. This means from expression (1.21b) that the component a_3 of the vector field **a** must be zero, $a_3 = 0$. In terms of the system (3.256), whenever appears something like

$$\frac{a_1}{b_1} = \frac{a_2}{b_2} = \frac{a_3}{0}\ , \tag{1.25}$$

this means that $a_3 = 0$ (not singularity of the system).

1.3 Methods of Solution of $dx/P = dy/Q = dz/R$

We pointed out in the last section that the integral curves of the set of differential equations

$$\frac{dx}{P} = \frac{dy}{Q} = \frac{dz}{R} \tag{1.26}$$

form a two-parameter family of curves in three dimensional space. Suppose we are able to derive from Eq. (1.26) two relations of the form

$$u_1(x,y,z) = c_1 \quad \text{and} \quad u_2(x,y,z), \tag{1.27}$$

where c_1 and c_2 are the constants of integration, then by varying these constants, we obtained a two-parameter family of curves satisfying the differential equations (1.26).

METHOD (I). Since any tangential direction (dx, dy, dz) at the point (x,y,z) on the surface $u_1(x,y,z) = c_1$ satisfies the relation

$$\frac{\partial u_1}{\partial x}dx + \frac{\partial u_1}{\partial y}dy + \frac{\partial u_1}{\partial z}dz = 0, \tag{1.28}$$

and according with the relation (1.19) we also have

$$\frac{\partial u_1}{\partial x}P + \frac{\partial u_1}{\partial y}Q + \frac{\partial u_1}{\partial z}R = 0. \tag{1.29}$$

To find u_1, we look for functions P', Q' and R' such that

$$P'P + Q'Q + Z'Z = 0, \tag{1.30}$$

i.e. a vector field $\mathbf{E'} = (P', Q', R')$ which is perpendicular to $\mathbf{E} = (P, Q, R)$ at every point (x,y,z). Because of Eq. (1.29), this vector satisfies

$$P' = \frac{\partial u_1}{\partial x}, \quad Q' = \frac{\partial u_1}{\partial y}, \quad R' = \frac{\partial u_1}{\partial z}. \tag{1.31}$$

Then, with (1.28) we would have that

$$P'dx + Q'dy + Z'dz$$

can be an exact differential, du_1 (if the integration can be done). The same procedure can be followed to obtain the other family of curves u_2.

Example 1.2. Find the integral curves of the equation

$$\frac{dx}{x(y-z)} = \frac{dy}{y(x-x)} = \frac{dz}{z(x-y)}$$

the vector field **E** is given as

$$\mathbf{E} = (x(y-z), y(x-x), z(x-y)).$$

If we take the vector fields $\mathbf{E}' = (1,1,1)$ and $\mathbf{E}'' = (zy, zx, xy)$ the condition (1.30) is satisfied and the functions u_1, u_2 of the equation (1.31) are

$$u_1 = x+y+z, \qquad u_2 = xyz$$

hence, the integral curves of the given differential equations are the members of the two-parameter family

$$x+y+z = c_1, \qquad xyz = c_2.$$

We must note that this method depends very much on the intuition and on the skill to determine the form of the vectors fields $\mathbf{E}'$, $\mathbf{E}''$.

METHOD (II). Suppose we find two vectors fields $\mathbf{E}' = (P', Q', R')$ and $\mathbf{E}'' = (P'', Q'', R'')$ such that the differentials

$$dw' = \frac{P'dx + Q'dy + Z'dz}{PP' + QQ' + RR'} \tag{1.32}$$

and

$$dw'' = \frac{P''dx + Q''dy + Z''dz}{PP' + QQ' + RR'} \tag{1.33}$$

are exact and are equal to each other. Then it follows

$$w' = w'' + c_1 \tag{1.34}$$

where c_1 is the integration constant.

Let us choose the vector $\mathbf{E}'$ of the form $\mathbf{E}' = (\lambda, \mu, \nu)$, where λ, μ and ν are constant and find the conditions that they must satisfy in order for the differential form (1.32), given by

$$\frac{1}{\rho} \frac{\lambda dx + \mu dy + \nu dz}{\lambda x + \mu y + \nu z}, \tag{1.35}$$

to be an exact differential, where ρ is another constant. From Eq. (1.26), this is possible only if the determinant of the matrix in the expression

$$\begin{pmatrix} -\rho & b & 1 \\ 1 & -\rho & c \\ a & 1 & -\rho \end{pmatrix} \begin{pmatrix} \lambda \\ \mu \\ \nu \end{pmatrix} = \begin{pmatrix} 0 \\ 0 \\ 0 \end{pmatrix} \tag{1.36}$$

is zero, that is if ρ is a root of the equation

$$-\rho^3 + (a+b+c)\rho + 1 + abc = 0.$$

This equation has three complex roots ρ_i, $i = 1,2,3$ and for each of them there exists a vector

$$\begin{pmatrix} \lambda_i \\ \mu_i \\ \nu_i \end{pmatrix} \qquad i = 1,2,3$$

which satisfies Eq. (1.36), and thus we have with Eq. (1.35) three possible exact differentials

$$dW' = d\log(\lambda_1 x + \mu_1 y + \nu_1 z)^{1/\rho_1},$$

$$dW'' = d\log(\lambda_2 x + \mu_2 y + \nu_2 z)^{1/\rho_2},$$

and

$$dW''' = d\log(\lambda_3 x + \mu_3 y + \nu_3 z)^{1/\rho_3}.$$

According to Eq. (1.34), we have the integral curves

$$(\lambda_1 x + \mu_1 y + \nu_1 z)^{1/\rho_1} = c_1(\lambda_2 x + \mu_2 y + \nu_2 z)^{1/\rho_2}$$

and

$$(\lambda_1 x + \mu_1 y + \nu_1 z)^{1/\rho_1} = c_2(\lambda_3 x + \mu_3 y + \nu_3 z)^{1/\rho_3}.$$

This method depends also on the intuition in determining the form of the vector fields $\mathbf{E}'$, $\mathbf{E}''$.

METHOD (III). When one of the variables is absent from one the equations of the set (1.19), it is possible to make a partial separation of variables, an we can derive the integral curves in a simple way. Suppose that the equation

$$\frac{dy}{Q} = \frac{dz}{R},$$

can be written in the form

$$\frac{dy}{dz} = f(y,z)$$

this equation has a solution of the form

$$\phi_1(y, z, c_1) = 0,$$

where c_1 is the integration constant. Solving this equation for z ($z = \psi(y, c_1)$) and substituting this value in the equation

$$\frac{dx}{P} = \frac{dy}{Q},$$

we obtain an ordinary differential equation of the type

$$\frac{dy}{dx} = g(x, y, c_1),$$

whose solution

$$\phi_2(x, y, c_1, c_2) = 0$$

may be readily obtained.

Example 1.3. Find the integral curves of the equations

$$\frac{dx}{ye^{-z}} = \frac{dy}{xe^y} = \frac{dz}{y/z}. \tag{1.37}$$

Using the first and third terms, we obtained the ordinary differential equation

$$\frac{dx}{dz} = xe^{-z},$$

which has the solution

$$x = c_1 e^{-e^{-z}}.$$

Substituting this in the second and third terms of (1.37), we obtain the ordinary differential equation

$$\frac{dy}{dz} = c_1^2 \frac{e^y}{y} e^{-2e^{-z}}.$$

This equation has the two-parametric solution

$$(y + 1)e^{-y} + c_1^2 \int e^{-2e^{-z}} = c_2.$$

1.4   Orthogonal Trajectories of a System of Curves on a Surface

The problem can be set as follows. Given a surface S defined by

$$F(x, y, z) = 0 \qquad (1.38a)$$

and a system of curves Γ on the surface formed by the set of relations (1.38a) and

$$G(x, y, z) = c_1, \qquad (1.38b)$$

find a system of curves $\{\Gamma'\}$ each of which lies on the surface (1.38a) and cuts every curve of the given system at right angles (see Fig. 1.10). The new system of curves is called the system of "orthogonal trajectories" on the surface of the given system of curves. Because the curves Γ are formed

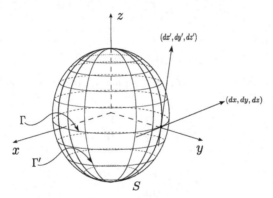

Fig. 1.10   System of orthogonal trajectories on the sphere an their tangent vectors.

from the intersection of surface (1.38a) and the family of surfaces (1.38b), the tangential direction of the curves Γ given by $(dx, dy, dz) = d\vec{\xi}$, at the point, (x, y, z) is perpendicular to the gradients of these surface

$$\nabla F \cdot d\vec{\xi} = \nabla G \cdot \vec{\xi} = 0 \qquad (1.39)$$

at the point, or

$$\frac{\partial F}{\partial x} dx = \frac{\partial F}{\partial y} dy = \frac{\partial F}{\partial z} dz = 0 \qquad (1.40)$$

and

$$\frac{\partial G}{\partial x} dx = \frac{\partial G}{\partial y} dy = \frac{\partial G}{\partial z} dz = 0. \qquad (1.41)$$

Hence, the vector $d\vec{\xi} = (dx, dy, dz)$ is parallel to the vector $\nabla F \times \nabla G$ (see Fig. 1.8) and must be such that

$$\frac{dx}{P} = \frac{dy}{Q} = \frac{dz}{R}, \tag{1.42}$$

where P, Q, and R are defined as

$$P = \frac{\partial(F,G)}{\partial(y,z)}, Q = \frac{\partial(F,G)}{\partial(z,x)}, R = \frac{\partial(F,G)}{\partial(x,y)}. \tag{1.43}$$

The tangent direction of the curves Γ', given by $d\vec{\xi'} = (dx', dy', dz')$ at the same point (x, y, z) is perpendicular to the gradient of the surface (1.38a) and must be perpendicular to the vector field $\mathbf{E} = (P, Q, R)$ (i.e $\nabla F \cdot d\vec{\xi'} = \mathbf{E} \cdot d\vec{\xi'} = 0$), then we have

$$\frac{\partial F}{\partial x}dx' = \frac{\partial F}{\partial y}dy' = \frac{\partial F}{\partial z}dz' = 0 \tag{1.44}$$

and

$$P dx' = Q dy' = R dz' = 0. \tag{1.45}$$

Hence, the vector $d\vec{\xi'} = (dx', dy', dz')$ is parallel to the vector $\nabla F \times \mathbf{E}$ and must be such that

$$\frac{dx'}{P'} = \frac{dy'}{Q'} = \frac{dz'}{R'}, \tag{1.46}$$

where

$$P' = R\frac{\partial F}{\partial y} - Q\frac{\partial F}{\partial z}, \tag{1.47a}$$

$$Q' = P\frac{\partial F}{\partial z} - R\frac{\partial F}{\partial x}, \tag{1.47b}$$

$$R' = Q\frac{\partial F}{\partial x} - P\frac{\partial F}{\partial y}. \tag{1.47c}$$

The solution of Eq. (1.46) together with the relation (1.38a) give the system of orthogonal trajectories Γ'.

Example 1.4. Find the orthogonal trajectories on the sphere $x^2 + y^2 + z^2 = a^2$ of its intersection with the family the of planes

$$z = k, \quad -a \le k \le a.$$

We have the functions

$$F = x^2 + y^2 + z^2 - a^2 = 0$$

and

$$G = z - k = 0,$$

the functions (1.43) are $P = 2y$, $Q = -2x$, and $R = 0$. The equations for the curve Γ are

$$\frac{dx}{2y} = \frac{dy}{-2x} = \frac{dz}{0},$$

which have solutions

$$x^2 + y^2 = a^2 - k^2.$$

The functions (1.47) are $P' = -4xz$, $Q' = 4yz$ and $R' = -4(x^2 + y^2)$. The system $\{\Gamma'\}$ of orthogonal trajectories to Γ on the sphere is determined by the equations

$$\frac{dx}{-4xz} = \frac{dy}{-4yz} = \frac{dz}{-4(x^2 + y^2)}$$

and

$$x^2 + y^2 = a^2 - z^2.$$

The solutions can be written as

$$xy = c_1, \qquad y^2(x^2 + y^2) = c_2.$$

1.5   Pfaffian Differential Equation in $\mathbb{R}^3$

The expression

$$w = Pdx + Qdy + Rdz, \tag{1.48}$$

where $(P, Q, R) = \mathbf{E}$ form a vector field in the space, is called a Pfaffian differential form a 1-form defined in $\mathbb{R}^3$. The equation generated from Eq. (1.48) (i.e. $w = 0$) as

$$Pdx + Qdy + Rdz = 0 \tag{1.49}$$

is called Pfaffian differential equation. This equation has the following geometric interpretation; let $U(x, y, z) = c$ a family of surfaces orthogonal to the vector field $\mathbf{E}$ and let $d\vec{\xi} = (dx, dy, dz)$ be the tangent direction to a given curve on this surface at the point (x, y, z). The gradient of the function U (∇U) is then parallel to the vector field $\mathbf{E}$ and thus, equation (1.49)

$$\mathbf{E} \cdot d\vec{\xi} = 0$$

would represent the family of surfaces perpendicular to the vector field $\mathbf{E}$.

If there exists such a family of surfaces, $U(x, y, z) = c$, orthogonal to the vector field $\mathbf{E}$, we say that the Pfaffian differential equation is "integrable". If there is not such a family, we say that the Pfaffian differential equation is "not integrable". If the Pfaffian is integrable, it means that there exists a function $\mu = \mu(x, y, z)$ such that $\mu\mathbf{E} = \nabla U$ or

$$\mu P = \frac{\partial U}{\partial x}, \quad \mu Q = \frac{\partial U}{\partial y}, \quad \mu R = \frac{\partial U}{\partial z} \tag{1.50}$$

and the function U is obtained through a line integration

$$U = \int_{(x_0, y_0, z_0)}^{(x, y, z)} \mu(P dx + Q dy + R dz), \tag{1.51}$$

if $\mu = 1$, we say that the vector field $\mathbf{E}$ is derivable from a "potential function U". We shall see that a necessary condition for the Pfaffian to be integrable is that $\mathbf{E} \cdot (\nabla \times \mathbf{E}) = 0$. In fact, if the Pfaffian (1.49) is integrable, the relations (1.50) are satisfied and assuming U is twice differentiable with $U_{xy} = U_{yx}$, $U_{yz} = U_{zy}$, and $U_{xz} = U_{zx}$, we have

$$\frac{\partial(\mu P)}{\partial y} = \frac{\partial(\mu Q)}{\partial x}, \frac{\partial(\mu Q)}{\partial z} = \frac{\partial(\mu R)}{\partial y}, \quad \text{and} \quad \frac{\partial(\mu R)}{\partial x} = \frac{\partial(\mu P)}{\partial z} \tag{1.52}$$

and as result of this differentiation, it follows

$$\mu\left(\frac{\partial P}{\partial y} - \frac{\partial Q}{\partial x}\right) = Q\frac{\partial \mu}{\partial x} - P\frac{\partial Q}{\partial y}, \tag{1.53a}$$

$$\mu\left(\frac{\partial Q}{\partial z} - \frac{\partial R}{\partial y}\right) = R\frac{\partial \mu}{\partial y} - Q\frac{\partial Q}{\partial z} \tag{1.53b}$$

and

$$\mu\left(\frac{\partial R}{\partial x} - \frac{\partial P}{\partial z}\right) = P\frac{\partial \mu}{\partial z} - R\frac{\partial Q}{\partial x}. \tag{1.53c}$$

Multiplying Eq. (1.53a) by R, Eq.(1.53b) by P, Eq. (1.53c) by Q, and adding them, we get

$$R\left(\frac{\partial P}{\partial y} - \frac{\partial Q}{\partial x}\right) + P\left(\frac{\partial Q}{\partial z} - \frac{\partial R}{\partial y}\right) + Q\left(\frac{\partial R}{\partial x} - \frac{\partial P}{\partial z}\right) = 0 \tag{1.54}$$

which is precisely the expressions $\mathbf{E} \cdot (\nabla \times \mathbf{E}) = 0$. It happens that is expressions is sufficient to assure that the Pfaffian (1.49) is integrable. We will not demonstrate this fact here. Once we know that the Pfaffian is integrable, we can use the relation (1.50) to find the family of surfaces orthogonal to the vector field $\mathbf{E}$.

Example 1.5. Verify that the differential equation

$$ydz + xdy + 2zdz = 0 \qquad (1.55)$$

is integrable and find its primitive. In this case

$$\mathbf{E} = (y, x, 2z) \qquad (1.56)$$

so that $\nabla \times \mathbf{E} = (0, 0, 0)$ and it is clear that $\mathbf{E} \cdot (\nabla \times \mathbf{E}) = 0$ and then $\mu = 1$. From Eq. (1.50) and Eq. (1.56), we have

$$y = \frac{\partial U}{\partial x}, \quad x = \frac{\partial U}{\partial y}, \quad \text{and} \quad 2z = \frac{\partial U}{\partial z}$$

upon integration, the solution is

$$U = xy + z^2. \qquad (1.57)$$

An alternative way to find the solution is to notice from (1.55) that it can be expressed as the exact differential

$$d(xy + z^2) = 0$$

so that solution (1.57) is again obtained.

1.6   Newton's Mechanics, Lagrangians, Hamiltonian, Hamilton-Jacobi Equation, and Liouville's Theorem

N-dimensional Newton's mechanics can be seen as a particular dynamical system described by the set the equations

$$\frac{dv_i}{dt} = f_i(\mathbf{q}, \mathbf{v}, t), \quad i = 1, ..., N \qquad (1.58a)$$

and

$$\frac{dq_i}{dt} = v_i, \quad i = i, ..., N, \qquad (1.58b)$$

where $\mathbf{q} = (q_1, ..., q_N)$, $\mathbf{v} = (v_1, ..., v_N)$ and "t" represented the generalized coordinates, generalized components of the velocity, and the time parameter respectively. f_i represents the total force acting on the particle in the direction "i" divided by the constant mass of the particle. Eq. (1.58a) and (1.58b) describe the motion of a single particle moving in a N-dimensional space under given force. These equations can be written in the same form as the relation (1.19)

$$\frac{dq_1}{dv_1} = ... = \frac{dq_N}{dv_N} = \frac{dv_1}{df_1} = ... = \frac{dv_N}{df_N} = dt. \qquad (1.59a)$$

The solution equations (1.58) or (1.59a) represent curves (or trajectories) in a $2N$-dimensional space $(\mathbf{q}, \mathbf{v})$ whose tangent directions are parallel to the vector field

$$\mathbf{E} = (v_1, ..., v_N, f_1, ..., f_N). \tag{1.59b}$$

For most of the physical problems, it is possible to deduce Eq. (1.58a) and Eq. (1.58b) from a variational principle (see reference of J. Douglas for the case $N = 2$),

$$\delta \int_a^b L dt = 0, \tag{1.60}$$

where L is called the Lagrangian of the system (1.58) and is a function depending on the generalized coordinates, velocities, and time

$$L = L(\mathbf{q}, \dot{\mathbf{q}}, t) \tag{1.61}$$

where $\dot{q}_i$ can be the ith-component of the velocity, $\dot{q}_i = v_i$. As a result of the variation of Eq. (1.60), we obtained the Euler equations (see reference of H. Goldstein for more details)

$$\frac{d}{dt}\left(\frac{\partial L}{\partial \dot{q}_i}\right) - \frac{\partial L}{\partial q_i} = 0, \quad i = 1, ..., N. \tag{1.62}$$

The usual approach in physical problems is starting with a more or less understood Lagrangian (1.61) to deduce the equations (1.58) through its differentiation (1.62). The inverse problem, however, starts with Eq. (1.58a) and Eq. (1.58b) to deduce the Lagrangian (1.61), this procedure is not so simple and is called "The Inverse Problem of the Calculus of Variations". Once we have the Lagrangian (1.61), we define a new variable p_i (called the generalized linear momentum) as

$$p_i = \frac{\partial L}{\partial \dot{q}_i} \tag{1.63}$$

and make a Legendre Transformation,

$$H = \sum_{i=1}^N \dot{q}_i p_i - L, \tag{1.64}$$

to define the Hamiltonian, $H = H(\mathbf{q}, \mathbf{p}, t)$, of the system (1.58). Through this procedure, it is assumed that we can have $\dot{q} = \dot{q}(\mathbf{q}, \mathbf{p}, t)$ to be able to substitute this on the right hand side of Eq. (1.64). Using Eq. (1.63) and Eq. (1.62), we get

$$dH = \sum_{i=1}^N \left(\frac{\partial H}{\partial q_i}dq_i + \frac{\partial H}{\partial p_i}dp_i\right) + \frac{\partial H}{\partial t}dt = \sum_{i=1}^N (\dot{q}_i dp_i - \dot{p}_i dq_i) - \frac{\partial L}{\partial t}.$$

Equaling coefficients and using the fact that the variables q_i and p_i are independent $i = 1, ..., N$, we obtain the Hamilton equations of motion

$$\frac{\partial q_i}{\partial t} = \frac{\partial H}{\partial p_i}, \qquad i = 1, ..., N \tag{1.65a}$$

$$\frac{\partial p_i}{\partial t} = -\frac{\partial H}{\partial q_i}, \qquad i = 1, ..., N \tag{1.65b}$$

and

$$\frac{\partial L}{\partial t} = -\frac{\partial H}{\partial t}. \tag{1.65c}$$

Note that Eq. (1.65a) and Eq. (1.65b) can be written in the form (1.19) as

$$\frac{dq_1}{\left(\dfrac{\partial H}{\partial p_1}\right)} = ... = \frac{dq_N}{\left(\dfrac{\partial H}{\partial p_N}\right)} = \frac{dp_1}{-\left(\dfrac{\partial H}{\partial q_1}\right)} = ... = \frac{dp_N}{-\left(\dfrac{\partial H}{\partial q_N}\right)}. \tag{1.66}$$

In this way, the motion of the particle in N-dimensions can be described by equations (1.58) or by equations (1.65). However, Eq. (1.65a) and Eq. (1.65b) give us certain advantages to observe the dynamics of the system. The solutions of equation (1.65) or (1.66) given us curves in a $2N$-dimensional space $(\mathbf{q}, \mathbf{p})$ called "Phase Space" whose tangent directions are parallel to the vector field

$$\mathbf{E} = \left(\frac{\partial H}{\partial p_1}, ..., \frac{\partial H}{\partial p_N}, -\frac{\partial H}{\partial q_1}, ..., -\frac{\partial H}{\partial q_N}\right).$$

We must notice that the existence of the Hamiltonian for system (1.58) depends on the existence of the Lagrangian, through the Legendre transformation (1.64) (see reference of D. Darboux for $N = 1$ and the reference of J. Douglas for $N = 2$), and given the Hamiltonian, the problem of the direct determination of the Lagrangian may be made using Eq. (1.63) and Eq. (1.64) through the solution of the nonlinear (in general) partial differential equation of first order

$$H\left(\mathbf{q}, \frac{\partial L}{\partial \dot{\mathbf{q}}}, t\right) - \sum_{i=1}^{N} \dot{q} \frac{\partial L}{\partial \dot{q}} + L = 0. \tag{1.67}$$

This expression represents a nonlinear partial differential equation of first order for the function L. Now, because the variables q_i and p_i, $i = 1, ..., N$ are independent, we can do, in general, a change in variable in the phase space of the form

$$Q_i = Q_i(\mathbf{q}, \mathbf{p}, t), \quad P_i = P_i(\mathbf{q}, \mathbf{p}, t) \quad i = 1, ..., N \tag{1.68}$$

satisfying, of course, the condition that the Jacobian of the transformation be different from zero,

$$\frac{\partial(Q, P)}{\partial(q, p)} \neq 0$$

(if $Q_i = Q_i(\mathbf{q}, t)$ and $P_i = P_i(\mathbf{q}, t)$, $i = 1, ..., N$ the transformation is called a point transformation). If under transformation (1.68), we have a function $\widetilde{V} = \widetilde{V}(Q, P, t)$ such that

$$\dot{Q}_i = \frac{\partial \widetilde{V}}{\partial P_i}, \quad \dot{P}_i = \frac{\partial \widetilde{V}}{\partial Q_i} \quad i = 1, ..., N \quad \text{and} \quad \frac{\partial \widetilde{L}}{\partial t} = -\frac{\partial \widetilde{V}}{\partial t}, \tag{1.69}$$

that is, the Hamilton equations (1.65) remain invariant, then the transformation (1.68) is called "canonical". It can be demonstrated that this happens when the Jacobian of the transformation is equal to 1, that is, the volume of the phase space is conserved.

Suppose that the Eq. (1.68) represents a canonical transformation. Since it satisfies Eq. (1.69), we can form the Lagrangian $\widetilde{L} = \widetilde{L}(\mathbf{Q}, \dot{\mathbf{Q}}, t)$ with a Legendre transformation like Eq. (1.64)

$$\widetilde{L} = \sum_{i=1}^{N} \dot{Q}_i P_i - \widetilde{V} \tag{1.70}$$

which will satisfy a similar principle to Eq. (1.60), given as

$$\delta \int_a^b \widetilde{L} dt = 0, \tag{1.71}$$

but since both of them describe te same system (1.58), they must differ a total time derivative of a function F, that is

$$L = \widetilde{L} + \frac{dF}{dt}. \tag{1.72}$$

This function F is called the "generating function" and once this one is given, the transformations (1.68) are completely specified. F makes the connection of the old variables $(\mathbf{q}, \mathbf{p})$ and the new ones $(\mathbf{Q}, \mathbf{P})$, thus F is a function of the old and new variables. There are four possibilities $F_1(\mathbf{q}, \mathbf{Q}, t)$, $S(\mathbf{q}, \mathbf{P}, t)$, $F_3(\mathbf{p}, \mathbf{Q}, t)$, and $F_4(\mathbf{p}, \mathbf{P}, t)$. Let us consider the second possibility, from Eq. (1.64), Eq. (1.70) and Eq. (1.72), using the fact $S(\mathbf{q}, \mathbf{P}, t) = F_1(\mathbf{q}, \mathbf{Q}, t) + \sum_{i=1}^{N} P_i Q_i$ and taking the total time differentiation of F_1, we obtain

$$\sum_{i=1}^{N} \left(p_i - \frac{\partial S}{\partial q_i} \right) \dot{q}_i + \sum_{i=1}^{N} \left(Q_i - \frac{\partial S}{\partial P_i} \right) \dot{P}_i + \widetilde{V} - H + \frac{\partial S}{\partial t} = 0,$$

and, therefore, we get the equations

$$p_i = \frac{\partial S}{\partial q_i}, \quad i = 1, ..., N, \tag{1.73a}$$

$$Q_i = \frac{\partial S}{\partial P_i}, \quad i = 1, ..., N \tag{1.73b}$$

and

$$\widetilde{V} = H + \frac{\partial S}{\partial t}. \tag{1.74}$$

Let us assume now that the transformation generated is such that $\dot{Q}_i = 0$ and $\dot{P}_i = 0$, $i = 1, ..., N$ which implies

$$Q_i = \beta_i, \quad P_i = \alpha_i \quad \text{(constants)} \quad i = 1, ..., N \tag{1.75}$$

and from Eq. (1.69), we can make $\widetilde{V} \equiv 0$, then from Eq. (1.74), we have

$$H\left(\mathbf{q}, \frac{\partial S}{\partial \mathbf{q}}, t\right) + \frac{\partial S}{\partial t} = 0. \tag{1.76}$$

Because of the relation (1.75), we have

$$S = S(\mathbf{q}, \alpha, t) \tag{1.77}$$

and

$$\beta_i = \frac{\partial S}{\partial \alpha_i} \quad i = 1, ..., N. \tag{1.78}$$

Eq. (1.76) is called "The Hamilton-Jacobi equation" associated to the system (1.58). Solving this equation and using Eq. (1.73a) and (1.73b), we can solve our physical problem. S is called "The Hamilton principal function" whose existence depends on the Hamiltonian through the relation (1.74). This expression represents a nonlinear partial differential equation of first order for the function $S(\mathbf{q}, \alpha, t)$.

In this way there are three possible ways to solve a physical problem in Classical Mechanics using Eq. (1.58) (Newton's equations), Eq. (1.65) (Hamilton's equations) or Eq. (1.76) (Hamilton-Jacobi's equation). Consider now a system of N particles moving in a three dimensional space, then as a first approach we could use any of the above three ways to describe our system, but it is almost impossible to be able to do this with the method mentioned before (so called many bodies problem). An alternative way is to use statical mechanics where "the state of a particle" is represented as a point in the $6N$-dimensional phase space $(\mathbf{q}, \mathbf{p})$, and "the state of the whole system" is represented by density function $\rho(\mathbf{q}, \mathbf{p}, t)$ (number of particles per unit phase space volume, $d\widetilde{N}/dV$). This density is called "ensemble" or "representative set" of the system. To see the motion of ρ in the phase space we take the total time derivative of ρ

$$\frac{d\rho}{dt} = \sum_{i=1}^{3N} \left(\frac{\partial \rho}{\partial q_i} \dot{q}_i + \frac{\partial \rho}{\partial p_i} \dot{p}_i \right) + \frac{\partial \rho}{\partial t}.$$

Using Eqs. (1.65), we have

$$\frac{d\rho}{dt} = \sum_{i=1}^{3N} \left(\frac{\partial \rho}{\partial q_i} \frac{\partial H}{\partial p_i} - \frac{\partial \rho}{\partial p_i} \frac{\partial H}{\partial q_i} \right) + \frac{\partial \rho}{\partial t}. \tag{1.79}$$

Defining the poisson bracket of two functions $f = f(\mathbf{q}, \mathbf{p}, t)$ and $g = g(\mathbf{q}, \mathbf{p}, t)$ as

$$\{f, g\} = \sum_{i=1}^{3N} \left(\frac{\partial f}{\partial q_i} \frac{\partial g}{\partial p_i} - \frac{\partial f}{\partial p_i} \frac{\partial g}{\partial q_i} \right), \tag{1.80}$$

Eq. (1.79) can be written as

$$\frac{d\rho}{dt} = \{\rho, H\} + \frac{\partial \rho}{\partial t}. \tag{1.81}$$

If the number of points a fixed volume is constant (there is not annihilation or creation of points, sinks our sources) and the volume remains constant for a fixed number of particles (Liouville's Theorem), for a system in thermodynamic equilibrium where ρ does not depend explicitly on time ($\partial \rho / \partial t = 0$), we have the equation

$$\{\rho, H\} = 0. \tag{1.82}$$

For a system out of equilibrium, we have $\partial \rho / \partial t \neq 0$.

Given the Hamiltonian for the N-particle system in three dimensions, Eq. (1.81) or Eq. (1.82), represents a partial differential equation of first order for ρ.

Example 1.6. The Lagrangian for one a dimensional single particle in a dissipative medium, where the friction force is proportional to the speed squared, is given by

$$L = \frac{1}{2} m \dot{q}^2 \exp\left(2\alpha q/m\right), \tag{1.83}$$

where m is the mass of the particle and α is a friction coefficient. Find the equation of the motion in the form (1.58) and (1.65), and give the partial differential equations Eq. (1.67), Eq. (1.76), and Eq. (1.82).

Using the Euler's equations (1.62) for $N = 1$, we have

$$m\ddot{q} = -\alpha\dot{q}.$$

Then, the equations the motion in the form (1.58) are

$$\frac{dv}{dt} = -\frac{\alpha}{m}v^2, \tag{1.84a}$$

$$\frac{dq}{dt} = v. \tag{1.84b}$$

Using Eq. (1.63), the generalized linear momentum is

$$p = m\dot{q}\exp\left(2\alpha q/m\right), \tag{1.85}$$

and substituting this in the Legendre transformation, Eq. (1.64), the Hamiltonian is given by

$$H = \frac{p^2}{2m}\exp\left(-2\alpha q/m\right). \tag{1.86}$$

Thus, Hamilton's equations of motion are

$$\frac{dq}{dt} = \frac{p}{m}\exp\left(-2\alpha q/m\right) \tag{1.87a}$$

and

$$\frac{dp}{dt} = \frac{\alpha p^2}{m^2}\exp\left(-2\alpha q/m\right). \tag{1.87b}$$

With Eq. (1.86) and Eq. (1.76), the Hamilton-Jacobi equation is given by

$$\frac{1}{2m}\exp\left(-2\alpha q/m\right)\left(\frac{\partial S}{\partial q}\right)^2 + \frac{\partial S}{\partial t} = 0, \tag{1.88}$$

and using Eq. (1.82), the Liouville's theorem is expressed as

$$\frac{\partial\rho}{\partial q} + \frac{\alpha p}{m}\frac{\partial\rho}{\partial p} + \frac{\partial\rho}{\partial t} = 0. \tag{1.89}$$

Finally, given the Hamiltonian (1.86), one way to find the Lagrangian (1.83) directly is through the solution of Eq. (1.67), i.e.

$$\frac{\exp\left(-2\alpha q/m\right)}{2m}\left(\frac{\partial L}{\partial\dot{q}}\right)^2 - \dot{q}\frac{\partial L}{\partial\dot{q}} + L = 0, \tag{1.90}$$

and it is not so difficult to demonstrate that the Lagrangian (1.83) indeed satisfies Eq. (1.90).

Example 1.7. The dynamical equations of a physical pendulum of mass "m", length "l", and quadratic velocity dissipation can be written as

$$\frac{d\theta}{dt} = v \tag{1.91a}$$

and

$$\frac{dv}{dt} = -\omega_0^2 \sin\theta - \frac{\alpha}{m} v|v|. \tag{1.91b}$$

The function $K = (\theta, v)$ given by

$$K = \left[v^2 + \frac{2\omega_0^2}{1 + (2\alpha/m)^2} \left(\frac{2\alpha sig(v)}{m} \sin\theta - \cos\theta \right) \right] \frac{m \exp\left(2\alpha\theta sig(v)/m\right)}{2} \tag{1.92}$$

is a constant of motion since it satisfies the partial differential equation

$$v\frac{\partial K}{\partial \theta} - \left(\omega_0^2 \sin\theta + \frac{\alpha}{m} v|v| \right) \frac{\partial K}{\partial v} = 0. \tag{1.93}$$

The function $sig(v)$ and the constant ω_0 are defined as

$$sig(v) = \begin{cases} +1, & \text{if } v > 0 \\ -1, & \text{if } v < 0 \end{cases}$$

and

$$\omega_0 = \sqrt{l/g},$$

being g the local acceleration due to gravity. Properly, (1.92) is a constant of motion restricted for $v > 0$ or $v < 0$ since whenever there is a cross through $v = 0$, its value must be changed to bring about the spiral shrinking motion of the particle in the phase space. The Lagrangian of the system is given by

$$L(\theta, v) = \left[v^2 - \frac{2\omega_0^2}{1 + (2\alpha/m)^2} \left(\frac{2\alpha sig(v)}{m} \sin\theta - \cos\theta \right) \right] \frac{m \exp\left(\frac{2\alpha\theta sig(v)}{m}\right)}{2} \tag{1.94}$$

and generalized linear momentum by

$$p(\theta, v) = mv \exp\left(2\alpha\theta sig(v)/m\right). \tag{1.95}$$

From this expression, we get

$$v(\theta, p) = \frac{p}{m} \exp\left(-2\alpha\theta sig(v)/m\right) \tag{1.96}$$

and the Hamiltonian of the system is given by

$$H(\theta, p) = \frac{p^2}{2m} \exp\left(-2\alpha\theta sig(v)/m\right)$$
$$+ \frac{m\omega_0^2 \exp\left(2\alpha\theta sig(v)/m\right)}{1 + (2\alpha/m)^2} \left[\frac{2\alpha sig(v)}{m} \sin\theta - \cos\theta\right] = 0. \quad (1.97)$$

Then, the Hamilton-Jacobi equation associated to this system is written as

$$\frac{1}{2m}\left(\frac{\partial S}{\partial \theta}\right)^2 + \frac{\partial S}{\partial t} + \frac{m\omega_0^2 \exp\left(2\alpha\theta sig(v)/m\right)}{1 + (2\alpha/m)^2}\left[\frac{2\alpha sig(v)}{m}\sin\theta - \cos\theta\right] = 0.$$
$$(1.98)$$

1.6.1 *Equivalent Hamiltonians*

As we saw in section 1.6, two Lagrangians $L(\mathbf{q}, \dot{\mathbf{q}}, t)$ and $\widetilde{L}(\mathbf{Q}, \dot{\mathbf{Q}}, t)$ are equivalents if there is a function F_1, called generatrix function, such that $L = \widetilde{L} + dF_1/dt$. This generatrix function is of the form $F_1(\mathbf{q}, \mathbf{Q}, t)$ and is related with the other possible generatrix functions, $F_2(\mathbf{q}, \mathbf{P}, t)$, $F_3(\mathbf{p}, \mathbf{Q}, t)$, and $F_4(\mathbf{p}, \mathbf{P}, t)$ through a Legendre transformation. To give this generatrix function is the same as to give the canonical transformation which changes the Hamiltonian (Lagrangian) into another one. The transformed Hamiltonian, $\widetilde{H}$, is related with the initial Hamiltonian, H, in the form

$$\widetilde{H} = H + \frac{dF_i}{dt}, \quad (1.99)$$

independently on which generatrix function is used, $i = 1, 2, 3, 4$. Therefore, we can say that two Hamiltonian $H(\mathbf{q}, \mathbf{p}, t)$ and $\widetilde{H}(\mathbf{Q}, \mathbf{P}, t)$ are equivalents if the above relation is satisfied for any generatrix function. Note that the units of both Hamiltonians must be the same to be able to have this expression. Given the generatrix function F_i, there are differential relations with its complementary variables, for example, for $F_1(\mathbf{q}, \mathbf{Q}, t)$ its complementary variables are related by $p_j = \partial F_1/\partial q_j$ and $P_j = -\partial F_1/\partial Q_j$. So, the equivalence of the Hamiltonians brings about the following nonlinear partial differential equations

$$\widetilde{H}(\mathbf{Q}, -\nabla_Q F_1, t) = H(\mathbf{q}, \nabla_q F_1, t) + \frac{\partial F_1}{\partial t}, \quad (1.100)$$

$$\widetilde{H}(\nabla_P F_2, \mathbf{P}, t) = H(\mathbf{q}, \nabla_q F_2, t) + \frac{\partial F_2}{\partial t}, \quad (1.101)$$

$$\widetilde{H}(\mathbf{Q}, -\nabla_Q F_3, t) = H(-\nabla_p F_3, \mathbf{p}, t) + \frac{\partial F_3}{\partial t} , \qquad (1.102)$$

$$\widetilde{H}(\nabla_P F_4, \mathbf{P}, t) = H(-\nabla_p F_4, \mathbf{p}, t) + \frac{\partial F_4}{\partial t}, \qquad (1.103)$$

where ∇_ξ represents the gradient with respect the variable ξ. Because of the structure of these partial differential equations of first order, there may be a solution of separable variables (depending of the Hamiltonian expressions), for example $F_1(q, Q, t) = f_1(q, t) + g(Q, t)$. However, this type of solution defines only a point-like transformation. Therefore, it is necessary to look for different type of solution.

Example 1.8. We saw already on previous section that any Hamiltonian $H(\mathbf{q}, \mathbf{p}, t)$ is equivalent to the Hamiltonian $\widetilde{H} = 0$ if we choose the generatrix function $F_2(\mathbf{q}, \mathbf{P}, t) = S(\mathbf{q}, t)$ as the Hamilton-Jacobi function, with $\mathbf{P} = \vec{\alpha}$ (constant). Then, this generatrix function satisfies the Hamilton-Jacobi equation (Eq. (1.76) above)

$$H(\mathbf{q}, \nabla_q S, t) + \frac{\partial S}{\partial t} = 0 . \qquad (1.104)$$

Example 1.9. Write the possible equations that establish the equivalence between the harmonic oscillator, $H = p^2/2m + m\omega^2 q^2/2$, and the Hamiltonian $\widetilde{H} = \omega P$ with $q, p, P \in \mathbb{R}$.
From the above equation one has

$$-\omega \frac{\partial F_1}{\partial Q} = \frac{1}{2m} \left(\frac{\partial F_1}{\partial q} \right)^2 + \frac{1}{2} m\omega^2 q^2 + \frac{\partial F_1}{\partial t} \qquad (1.105)$$

$$\omega P = \frac{1}{2m} \left(\frac{\partial F_2}{\partial q} \right)^2 + \frac{1}{2} m\omega^2 q^2 + \frac{\partial F_2}{\partial t} \qquad (1.106)$$

$$-\omega \frac{\partial F_3}{\partial Q} = \frac{p^2}{2m} + \frac{1}{2} m\omega^2 \left(\frac{\partial F_3}{\partial p} \right)^2 + \frac{\partial F_3}{\partial t} \qquad (1.107)$$

$$\omega P = \frac{p^2}{2m} + \frac{1}{2} m\omega^2 \left(\frac{\partial F_4}{\partial p} \right)^2 + \frac{\partial F_4}{\partial t} \qquad (1.108)$$

which defines non linear partial differential equations in $\mathbb{R}^3$.

Of course, once we know a generatrix function, the other ones are determined through a Legendre transformation, for example, assume that we have determine the generatrix function $F_1(\mathbf{q}, \mathbf{Q}, t)$, then, since the generatrix function $F_2(\mathbf{q}, \mathbf{P}, t)$ is related with F_1 through the Legendre transformation $F_1(q, Q, t) = F_2(\mathbf{q}, \mathbf{P}, t) - \mathbf{Q} \cdot \mathbf{P}$, and since one has the relation

$Q_i = \partial F_2/\partial P_i$, we cant think this Legendre transformation as the following non linear partial differential equation of first order

$$\sum_i P_i \frac{\partial F_2}{\partial P_i} + F_1(\mathbf{q}, \nabla_P F_2, t) = F_2(\mathbf{q}, \mathbf{P}, t). \tag{1.109}$$

In the same way, assuming $F_1(\mathbf{q}, \mathbf{Q}, t)$ is known, one could have the generatrix functions F_3 and F_4 determined by the following non linear partial differential equations of first order

$$\sum_i p_i \frac{\partial F_3}{\partial p_i} + F_1(-\nabla_p F_3, \mathbf{Q}, t) = F_3(\mathbf{p}, \mathbf{Q}), t) \tag{1.110}$$

$$\sum_i \left\{ p_i \frac{\partial F_4}{\partial p_i} + P_i \frac{\partial F_4}{\partial P_i} \right\} + F_1(-\nabla_p F_4, \nabla_P F_4, t) = F_4(\mathbf{p}, \mathbf{P}, t). \tag{1.111}$$

Example 1.10. Let us assume that the generatrix function F_1 is given as

$$F_1(q, Q) = -\frac{m\omega^2 q^2}{2} \tan Q, \tag{1.112}$$

where m and ω are real parameters. This generatrix function establishes the equivalence between the Hamiltonian $H = p^2/2m + m\omega^2 x^2/2$ and the Hamiltonian $\widetilde{H} = \omega P$ since is a non separable solution of the equation $-\omega(\partial F_1/\partial Q) = (\partial F_1/\partial q)^2/2m + m\omega^2 q^2/2$. Then, according to what we said above, the generatrix function $F_2(q, P)$ can be found as the solution of the non linear partial differential equation

$$P \frac{\partial F_2}{\partial P} - \frac{m\omega^2 q^2}{2} \tan \left(\frac{\partial F_2}{\partial P} \right) = F_2(q, P). \tag{1.113}$$

Note from the Legendre transformation that there is always a variable which is not involved in the differentiation, in our last example this variable is "q." This fact also happens for the Legendre transformation between the Lagrangian, $L(\mathbf{q}, \dot{\mathbf{q}}, t)$, and the Hamiltonian, $H(\mathbf{q}, \mathbf{p}, t)$,

$$\sum_i \dot{q}_i \frac{\partial L}{\partial \dot{q}_i} - H(\mathbf{q}, \nabla_{\dot{q}} L, t) = L(\mathbf{q}, \dot{\mathbf{q}}, t). \tag{1.114}$$

In fact, if this Legendre transformation is though as a function of $\mathbf{q}$ and $\dot{\mathbf{q}}$ instead, and by calling $K(\mathbf{q}, \dot{\mathbf{q}}, t) = H(\mathbf{q}, \mathbf{p}(\mathbf{q}, \dot{\mathbf{q}}, t), t)$, one would have instead the following linear partial differential equation of first order

$$\sum_i \dot{q}_i \frac{\partial L}{\partial \dot{q}} = L(\mathbf{q}, \dot{\mathbf{q}}, t) + K(\mathbf{q}, \dot{\mathbf{q}}, t). \tag{1.115}$$

1.7   *Problems*

1.1   Find the tangent planes at the point $P = (0, \sqrt{3}, 1)$ of the surfaces $x^2 + y^2 + z^2 = 4$ and $z = 1$. Find the normal vectors to these tangent planes at the point, the line generated by intersection of these planes, and the curve generated by the intersection of both surfaces.

1.2   Find the integral curves of the follows equations, using the method (I) of Sec. 1.3

$$\frac{dx}{xz - y} = \frac{dy}{yz - x} = \frac{dz}{1 - z^2}.$$

1.3   Find the integral curves of the equations, using the method (II) of Sec. 1.3

$$\frac{dx}{y + az} = \frac{dy}{z + bx} = \frac{dz}{x + cy}.$$

1.4   Find the integral curves of the equations, using the method (II) of Sec. 1.3

$$\frac{adx}{(b - c)yz} = \frac{bdy}{(c - a)zx} = \frac{cdz}{(a - b)xy}.$$

1.5   Find the integral curves of the equations, using the method (III) of Sec. 1.3

$$\frac{dx}{x + z} = \frac{dy}{y} = \frac{dz}{z + y^2}.$$

1.6   Find the orthogonal trajectories on the cone $x^2 + y^2 = z^2 \tan^2 \alpha$ of its intersection with the family of planes parallel to $z = 0$.

1.7   Verify that the differential equation

$$(y^2 + yz)dz + (xz + z^2)dy + (y^2 - xy)dz = 0$$

is integrable and find its primitive.

1.8   Do the same as on the Exercise 6, but for the one dimensional oscillator whose Lagrangian is given by

$$L = \frac{1}{2}m\dot{q}^2 - \frac{1}{2}kq^2,$$

where m is the mass of the particle and k is the spring constant.

1.9   Write down the non-linear partial differential equation to find the Lagrangian directly if the Hamiltonian is given by (1.97).

1.10 Write the equation which would establish the equivalence between the Hamiltonians $H_1 = p^2/2m + k/q$ and $H_2 = p^2/2m + m\omega^2 q^2/2$ (select one Hamiltonian with variable Q and P).

1.11 Given the generatrix function $F_1 = \sum_{i=1}^{2}(\tan Q_i - \dot{\beta}_i(t))x_i^2/2\beta_i(t)$, write down the partial differential equation of first order which helps to find the generatrix function $F_2(x_1, x_2, P_1, P_2, t)$.

1.12 Given the Euler-Lagrange equation for one degreed of freedom,

$$\frac{dL_v}{dt} - L_x = 0 \ ,$$

where one has defined $L_v = \partial L/\partial v$ and $L_x = \partial L/\partial x$, make the differentiation of this expression with respect to "v," and show that a PDEFO with respect to $J = L_{vv}$ given by

$$v\frac{\partial J}{\partial x} + F(x,v)\frac{\partial J}{\partial v} = -F_v J \ ,$$

where $dx/dt = v$ and $dv/dt = F(x,v)$ is the defined autonomous system.

1.13 If an arbitrary conservative system with three degrees of freedom is characterized by the potential function $V(\mathbf{x})$, show (using Euler-Lagrange equations) that

$$L = \left(\frac{1}{2}mv^2 - V(\mathbf{x})\right)e^{\alpha t/m}$$

represents an explicitly time depending Lagrangian for the associated dissipative system

$$m\frac{d\mathbf{v}}{dt} = \mathbf{F}(\mathbf{x}) - \alpha\mathbf{v} \ ,$$

where $\mathbf{v}$ is the velocity of the particle with speed v and constant mass m, $\mathbf{F}(\mathbf{x}) = -\nabla V(\mathbf{x})$ is the conservative force, and α is a non negative real constant.

1.14 Show that the Hamiltonian and Hamiton equations for the previous exercise are given by

$$H = \frac{p^2}{2m}e^{-\alpha t/m} + V(\mathbf{x})e^{\alpha t/m}, \quad \dot{x}_i = \frac{p_i}{m}e^{-\alpha t/m}, \quad \dot{p}_i = -\frac{\partial V}{\partial x_i}e^{\alpha t/m},$$

$$(1.116)$$

and write the Hamilton-Jacobi equation associated to this problem.

1.8   *References*

1.1   I.N. Sneddon, *Elements of Partial Differential Equations*, Mc Graw Hill Books Company, Inc. 1957. Chap.I

1.2   F.H. Miller, *Partial Differential Equations*, John Wiley & Sons, Inc. 1941. Chap. II.

1.3   L. Elsgoltz, *Ecuaciones Diferenciales y Cálculo Variacional*, Ciencia 1975, Chap. 5.

1.4   M. Spivak, *Calculus on Manifolds*, W.A. Benjamin, Inc., 1965. Chap. 4.

1.5   H. Goldstein, *Classical Mechanics*, Addison-Wesley Publishing Co. Inc., 1965. Chap.1.

1.6   J. Douglas, Trans. Amer. Math. Soc. **50**, (1947) 71.

1.7   I.M. Gelfand and S.V. Fomins, *Calculus of Variations*, Prentice-Hall, Inc. ,1963. Chap. 1-5.

1.8   G. López, Ann. Phys., **251**, No.2 (1996) 372.

1.9   J.A. Kobussen, Acta Phys. Austr. **51** (1979) 293.

1.10   C. Leubner, Phys. Lett. A, **86** (1981) 2.

1.11   D. Darboux, *Leçons sur la théorie général des surfaces et les applications géométriques du calcul infinetésimal*, Gauthier-Villars, Paris, 1984. IViéme partie.

Chapter 2

Partial Differential Equations of First Order

In this chapter we make the introduction to the Partial Differential Equation of First Order (henceforth abbreviated as PDEFO) considering the classification and the solution of a particular kind of PDEFO.

2.1 Classification

A partial differential equation for a function z defined in a n-dimensional space is an arbitrary relation of the form

$$F(\mathbf{x}, z, z_i; z_{i_1 i_2}; ...; z_{i_1,...,i_m}) = 0 \qquad (2.1)$$

where $\mathbf{x} = (x_1, ..., x_n)$ is a point in the space $\mathbb{R}^n$, $z_i, z_{i_1 i_2}$ and $z_{i_1,...,i_m}$ are the partial derivatives defined as

$$z_i = \frac{\partial z}{\partial x_i} \quad i = 1, ..., n, \qquad (2.2a)$$

$$z_{i_1 i_2} = \frac{\partial^2 z}{\partial x_{i_1} \partial x_{i_2}} \quad i_1, i_2 = 1, ..., n, \qquad (2.2b)$$

and

$$z_{i_1,...,i_m} = \frac{\partial^m z}{\partial x_{i_1}, ..., \partial x_{i_m}} \quad i_j = 1, ..., n \qquad (2.2c)$$

where m is a natural number. We call $z = \phi(x_1, ..., x_n)$ a solution of (2.1) if after substitution of z and its partial derivatives, Eq. (2.1) is satisfied identically in every point $\mathbf{x}$ of some region $\Omega \subset \mathbb{R}^n$ of the space. We will assume also that the function z and its derivatives (2.2) up to the order $k \in Z$ are continuous in the region Ω that is, z is a class $C^k(\Omega)$, where $C^k(\Omega)$ denotes the set of real functions defined in Ω with its derivatives continuous up to the order k. It is not difficult to prove that $C^k(\Omega)$ forms a vector space over the real space $\mathbb{R}$. The order of the partial differential equation is the order of the highest derivative that occurs in (2.1), the natural number m.

A PDEFO for a function z defined in region Ω of the space $\mathbb{R}^n$, is a relation of the form

$$F\left(\mathbf{x}, z, \frac{\partial z}{\partial x_1}, ..., \frac{\partial z}{\partial x_n}\right) = g(\mathbf{x}, z), \tag{2.3}$$

where the pure dependence on the $\mathbf{x}$ and z has been separated through the function $g(\mathbf{x}, z)$. A solution $z = f(x_1, ..., x_n)$ of Eq. (2.3), when interpreted as a n-dimensional surface, will be called an "integral surface" of the differential equation.

The PDEFO (2.3) can be see, fixing the point $\mathbf{x} \in \Omega$, as a functional F acting in the space $C^k(\Omega)$ with values in the real space $\mathbb{R}$. According with this observation, we will say that a PDEFO is linear if for every $z_1, z_2 \in C^k(\Omega)$ and $\alpha \in \mathbb{R}$, we have

$$F\left(\mathbf{x}, z_1 + \alpha z_2, \frac{\partial (z_1 + \alpha z_2)}{\partial x_i}\right) = F\left(\mathbf{x}, z_1, \frac{\partial z_1}{\partial x_i}\right) + \alpha F\left(\mathbf{x}, z_2, \frac{\partial z_2}{\partial x_i}\right) \tag{2.4}$$

i.e., Eq. (2.3) is linear with respect to z and its partial derivatives. We will say that a PDEFO is quasi-linear if instead of the relation (2.4), we have

$$F\left(\mathbf{x}, z_1 + \alpha z_2, \frac{\partial (z_1 + \alpha z_2)}{\partial x_i}\right) = F_1\left(\mathbf{x}, z_1 + \alpha z_2, \frac{\partial z_1}{\partial x_i}\right)$$
$$+ \alpha F_2\left(\mathbf{x}, z_1 + \alpha z_2, \frac{\partial z_2}{\partial x_i}\right), \tag{2.5}$$

where F_1 and F_2 could be different functions. That is, Eq. (2.3) is only linear with respect to the partial derivatives. We will say that the PDEFO is non-linear if neither relation (2.4) or (2.5) is satisfied.

A general linear PDEFO can be written as

$$\sum_{i=1}^{n} a_i(\mathbf{x}) \frac{\partial z}{\partial x_i} - c(\mathbf{x}) z - d(\mathbf{x}) = 0, \tag{2.6}$$

and a general quasi-linear PDEFO can be written as

$$\sum_{i=1}^{n} a_i(\mathbf{x}, z)\frac{\partial z}{\partial x_i} - c(\mathbf{x}, z) = 0. \tag{2.7}$$

An example of a non-linear PDEFO is the following equation

$$\sum_{i=1}^{n} a_i(\mathbf{x})\left(\frac{\partial z}{\partial x_i}\right)^2 = 0. \tag{2.8}$$

2.2 Linear PDEFO for Functions Defined in $\Omega \subset \mathbb{R}^2$

The theory we will discuss in this section is extended immediately to any number of variables (dimensions). We restrict ourselves to two variables $(x, y) \in \Omega \subset \mathbb{R}^2$ to make clear the interesting geometric interpretation of the linear PDEFO. Eq. (2.6) can be written as

$$a(x, y)\frac{\partial z}{\partial x} + b(x, y)\frac{\partial z}{\partial y} = c(x, y)z + d(x, y). \tag{2.9}$$

We notice that the left hand side of this equation represents the derivative of $z(x, y)$ in the direction

$$\mathbf{E}(x, y) = (a(x, y), b(x, y))$$

that is,

$$a(x, y)\frac{\partial z}{\partial x} + b(x, y)\frac{\partial z}{\partial y} = \mathbf{E} \cdot \nabla z$$

thus, when we consider the curves in the $x - y$ plane whose tangents $(d\vec{\xi})$ at each point have those directions $(d\vec{\xi} \sim \mathbf{E})$; that is, the one parametric family of curves defined by the ordinary differential equations

$$\frac{dx}{dt} = a(x, y), \quad \frac{dy}{dt} = b(x, y) \tag{2.10a}$$

or the single equation

$$\frac{dx}{dy} = \frac{a(x, y)}{b(x, y)}, \tag{2.10b}$$

is such that along this curve, $z(x, y)$ will satisfy the ordinary differential equation

$$\frac{dz}{dx} = c(x,y)z + d(x,y) \qquad (2.11a)$$

or

$$\frac{dz}{dx} = \frac{c(x,y)z + d(x,y)}{a(x,y)}. \qquad (2.11b)$$

Let us prove this. Differentiating z with respect to the parameter t, we have

$$\frac{dz}{dt} = \frac{\partial z}{\partial x}\frac{dx}{dt} + \frac{\partial z}{\partial y}\frac{dy}{dt},$$

along the curves (2.10), this expression becomes

$$\frac{dz}{dt} = a(x,y)\frac{\partial z}{\partial x} + b(x,y)\frac{\partial z}{\partial y},$$

but according to Eq. (2.9), we get Eq. (2.11). The one parametric family of curves defined by Eq. (2.10) are called the "characteristic curves" of the differential equation, which will be denoted by $\{c_\lambda\}$.

Suppose now that $z(x,y)$ is assigned an "initial" value at the point x_0, y_0 in the $x - y$ plane. From the existence and uniqueness of the initial value problem for ordinary differential equations, Eq. (2.10) will define a unique characteristic curve, say

$$x = x(x_0, y_0, t), \quad y = y(x_0, y_0, t) \qquad (2.12)$$

along which

$$z = z(z_0, x_0, y_0, t) \qquad (2.13)$$

will be uniquely determined by Eq. (2.11). That is, If z is given at a point, it is determined along a whole characteristic curve through the point. This suggests that if we were to assign initial values for z along some curve Γ (Fig. 2.1), intersecting the characteristic c_λ, we may determine a unique solution $z(x,y)$ in the whole region covered by the family c_λ by means of Eq. (2.12) and Eq. (2.13).

We can anticipate problems with the determination of the solution $z(x,y)$ for curves like Γ' shown in Fig. 2.1 because for several points of the characteristic curve c_0, we have the same solution (2.13). The curve Γ, which

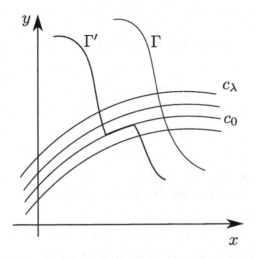

Fig. 2.1 The initial curve Γ determine an unique solution $z(x,y)$, meanwhile $z(x,y)$ is not unique for Γ'.

we may call the initial curve, may not be chosen quite arbitrary. Clearly, it must not coincide with the characteristic curve since z must be determined as a solution of an ordinary differential equation in unique form.

A precise formulation of this initial value problem, called the Cauchy initial value problem, will be given for the more general quasi-linear equations to follow. However, the next example points out some important feature.

Example 2.1. Find the integral surface of the equation

$$\frac{1}{y}\left(\frac{\partial z}{\partial x}\right) + \frac{\partial z}{\partial y} = \frac{e^{-x}}{x}z$$

with initial conditions $z = \phi(x)$ for $y = 1$. The equation for the characteristic curve is given by

$$\frac{dy}{dx} = y$$

having the solutions

$$y = ce^x.$$

Along such a curve, z satisfies Eq. (2.11b)

$$\frac{dz}{dx} = \frac{ye^{-x}}{x}z = \frac{c}{x}z$$

which has the solution

$$z = K(c)x^c.$$

The constant K depends on c because it may differ from characteristic to characteristic, then we have the general solution

$$z(x,y) = K(ye^{-x})x^{ye^{-x}}, \qquad (2.14)$$

where K is "arbitrary function". If we apply the initial condition for $y = 1$, we obtain

$$\phi(x) = K(e^{-x})x^{e^{-x}}$$

or

$$K(s) = \frac{\phi(\log s^{-1})}{(\log s^{-1})^s},$$

hence, the functionality $K(ye^{-x})$ is determined, and the required solution is

$$z(x,y) = \frac{\phi(x - \log y)x^{ye^{-x}}}{(x - \log y)^{ye^{-x}}}.$$

Let us see what happens if instead of having the above initial curve $y = 1$, we choose the characteristic curve given by $y = 2e^x$. With the initial condition $z = \phi(x)$ at $y = 2e^x$ substituted in (2.14), we cannot determine the functionality of K because we get the following relation

$$\phi(x) = K(2)x^2.$$

Thus, the Cauchy problem is not well posed on characteristic curves. We need to take care that our initial curve be different from any characteristic curve of the differential equation.

2.3   Quasi-Linear PDEFO for Functions Defined in $\Omega \subset \mathbb{R}^2$

According with (2.7), the general quasi-linear equation may be written as

$$P(x,y,z)\left(\frac{\partial z}{\partial x}\right) + Q(x,y,z)\left(\frac{\partial z}{\partial y}\right) = R(x,y,z). \qquad (2.15)$$

The solution $z = f(x,y)$ defines an integral surface in the three dimensional x, y, z space. As we saw in the relation (1.38), the directions numbers of the normal to the surface are $(\partial z/\partial x, \partial z/\partial y, -1)$, so that equation (2.15) can be interpreted as the condition for the integral surface at each point to have the property that the vector field $\mathbf{E} = (P, Q, R)$ is tangent to the surface. Thus, the quasi-linear equation (2.15) defines a direction field $\mathbf{E}$ called the characteristic direction, having the property that a surface $z = f(x,y)$ is an integral surface if and only if at each point(x, y, z) the tangent plane contains the characteristic direction (see Fig. 2.2).

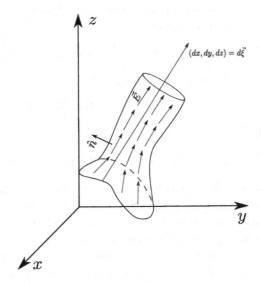

Fig. 2.2   For a surface $z = f(x,y)$ the field $\mathbf{E}$ gives us the characteristic directions.

It is suggestive then that we consider the integral curves of this field $\mathbf{E}$ i.e., the family of space curves whose tangent direction $d\vec{\xi} = (dx, dy, dz)$ coincides with the direction of the vector field $\mathbf{E}$ $(d\vec{\xi} \sim \mathbf{E})$. They are called

40   *Partial Differential Equations of First Order*

the characteristic curves and are given by equations

$$\frac{dx}{P(x,y,z)} = \frac{dy}{Q(x,y,z)} = \frac{dz}{R(x,y,z)}. \tag{2.16a}$$

Calling the common value of these ratios dt (the constant of proportionality between the vector $d\vec{\xi}$ and $\mathbf{E}$). We can write Eq. (2.16a) in the form

$$\frac{dx}{dt} = P(x,y,z), \quad \frac{dy}{dt} = Q(x,y,z), \quad \frac{dz}{dt} = R(x,y,z). \tag{2.16b}$$

This notion differs from the one used in the linear case. The projection of the present curves on the $x-y$ plane will be the curves previously called the characteristics. Through each point (x_0, y_0, z_0) passes one characteristic curve

$$x = x(x_0, y_0, z_0, t), \quad y = y(x_0, y_0, z_0, t), \quad z = z(x_0, y_0, z_0, t).$$

One important property of the characteristic curves is immediately evident from the geometric interpretation of equation (2.15). Namely, every surface generated by one parametric family of characteristic is an integral surface. Moreover, the converse is also true, if $z = f(x,y)$ is a given integral surface S. Consider the solution of

$$\frac{dx}{dt} = P(x, y, f(x,y)), \quad \frac{dy}{dt} = P(x, y, f(x,y))$$

with $x = x_0, y = y_0$ for $t = 0$. Then, for the corresponding curve

$$x = x(t), \quad y = y(t), \quad z = f(x(t), y(t)),$$

we have from Eq. (2.15)

$$\frac{dz}{dt} = f_x\frac{dx}{dt} + f_y\frac{dy}{dt} = P(x,y,z)f_x + Q(x,y,z)f_y = R(x,y,f).$$

Hence, the curve satisfies condition (2.16b) for characteristic curves, and also lies on S by definition. Thus, S contains together with each point also the characteristic curve through the point. Therefore, if two integral surfaces intersect at a point, they intersect along the whole characteristic through this point because of the uniqueness of the solution (see below); and the curve of intersection of two integral surfaces must be characteristic.

The integral surface is also called the vector surface because the tangent of the curves lying on it have the same direction of the vector field $\mathbf{E}$. If the integral surface is given implicitly of the form

$$F(x,y,z) = 0. \tag{2.17}$$

The normal vector $\hat{n}$ at any point of the surface is given by the normalization to unit of the vector

$$\nabla F = \left(\frac{\partial F}{\partial x}, \frac{\partial F}{\partial y}, \frac{\partial F}{\partial z} \right) \qquad (2.18)$$

and the integral surface is determined by the condition of the vector field **E** which is orthogonal to the normal vector of the surface (2.17)

$$\mathbf{E} \cdot \nabla F = P(x,y,z)\frac{\partial F}{\partial x} + Q(x,y,z)\frac{\partial F}{\partial y} + R(x,y,z)\frac{\partial F}{\partial z} = 0. \qquad (2.19)$$

The characteristic curves of Eq. (2.19) are determined by the same set of equations (2.16). We must note that Eq. (2.19) is a linear PDEFO with three variables (x, y, z).

Let us assume that we have two independent characteristic curves when solving Eq. (2.16a), given as

$$\psi_1(x,y,z) = c_1, \quad \psi_2(x,y,z) = c_2. \qquad (2.20)$$

These are solution of Eq. (2.19). To show this fact, let us take the differential of one of the solutions given by Eq. (2.20)

$$d\psi_1 = \frac{\partial \psi_1}{\partial x}dx + \frac{\partial \psi_1}{\partial y}dy + \frac{\partial \psi_1}{\partial z}dz = 0$$

but along any integral curve we have $d\vec{\xi} \sim \mathbf{E}$, thus we get

$$\frac{\partial \psi_1}{\partial x}P + \frac{\partial \psi_1}{\partial y}Q + \frac{\partial \psi_1}{\partial z}R = 0 ,$$

but then, it is clear that any function of both characteristic curves is also a solution

$$\Phi(\psi_1, \psi_2) = 0 , \qquad (2.21)$$

and it is equivalent to say that the two integration constants that appear in Eq. (2.20) are not independent (this fact used in example 2.1 of this chapter above).

Example 2.2 Find the two characteristic curves of the equation

$$z\frac{\partial z}{\partial x} + \frac{\partial z}{\partial y} = 1.$$

The equation of the characteristic curves are

$$\frac{dx}{z} = dy = dz$$

which have the solutions

$$\psi_1 = z - y = c_1$$

and

$$\psi_1 = z^2 - 2x = c_2.$$

Thus, the general integral surface of the equation can be expressed as

$$\Phi(c_1, c_2) = 0$$

or

$$z^2 - 2x = c_2(c_1) \, ,$$

where c_2 is arbitrary function of c_1 (or other way around).

The solution of the Cauchy initial problem will be given in the next theorem.

Theorem 2.1. *Consider the first order quasi-linear partial differential equation (2.15), where P, Q, R have continuous partial derivatives with respect x, y, z. Suppose that along the initial curve $x = x_0(s), y = y_0(s)$, the initial values $z = z_0(s)$ are prescribed, x_0, y_0, z_0 being continuously differentiable function for $0 \leq s \leq 1$. Furthermore, let these function satisfy the expression*

$$\frac{dy_0}{ds} P(x_0(s), y_0(s), z_0(s)) - \frac{dx_0}{ds} P(x_0(s), y_0(s), z_0(s)) \neq 0. \qquad (2.22)$$

Then, there exist one and only one solution $z(x, y)$ defined in some neighborhood of the initial curve, which satisfies the partial differential equation and the initial condition

$$z(x_0(s), y_0(s)) = z_0(s).$$

Proof. From the existence and uniqueness theorem for ordinary differential equations we may solve equations (2.16b) for a unique family of characteristic

$$x = x(x_0, y_0, z_0, t) = x(s, t), \qquad (2.23a)$$
$$y = y(x_0, y_0, z_0, t) = y(s, t), \qquad (2.23b)$$
$$z = z(x_0, y_0, z_0, t) = z(s, t) \qquad (2.23c)$$

whose derivatives with respect to the parameter s, t are continuous such that they satisfy the initial conditions

$$x(s, 0) = x_0(s), \quad y(s, 0) = y_0(s), \quad z(s, 0) = z_0(s).$$

We note that the Jacobian

$$\frac{\partial(x,y)}{\partial(s,t)}\bigg|_{t=0} = \det \begin{pmatrix} x_s & x_t \\ y_s & y_t \end{pmatrix}_{t=0} = \frac{dx_0}{ds}Q - \frac{dy_0}{ds}P \neq 0$$

because of the condition (2.22). Thus, in Eq. (2.23) we many solve for s, t in terms of x, y in the neighborhood of the initial curve $t = 0$, obtaining from Eq. (2.23) a candidate for solution

$$\phi(x,y) = z(s(x,y), t(x,y)). \tag{2.24}$$

This function satisfies the following initial conditions

$$\phi(x,y)_{t=0} = z(s,0) = z_0(s)$$

and satisfies also the differential equation (2.15) since one has

$$
\begin{aligned}
P\phi_x + Q\phi_y &= P\left(\frac{\partial z}{\partial s}s_x + \frac{\partial z}{\partial t}t_x\right) + Q\left(\frac{\partial z}{\partial s}s_y + \frac{\partial z}{\partial t}t_y\right) \\
&= \frac{\partial z}{\partial s}(Ps_x + Qs_y) + \frac{\partial z}{\partial t}(Pt_x + Qt_y) \\
&= \frac{\partial z}{\partial s}\left(\frac{dx}{dt}s_x + \frac{dy}{dt}s_y\right) + \frac{\partial z}{\partial t}\left(\frac{dx}{dt}t_x + \frac{dy}{dt}t_y\right) \\
&= \frac{\partial z}{\partial s}\left(\frac{ds}{dt}\right) + \frac{\partial z}{\partial t}\left(\frac{dt}{dt}\right) \\
&= \frac{\partial z}{\partial s}(0) + \frac{\partial z}{\partial t} \\
&= \frac{\partial z}{\partial t} \\
&= R(x,y,z).
\end{aligned}
$$

Moreover, $\phi(x,y)$ is unique. Suppose that $\phi'(x,y)$ is any other solution satisfying the initial condition and x', y' is an arbitrary point in the neighborhood of the initial curve. We consider the characteristic curve on the surface ϕ

$$x = x(s',t'), \quad y = y(s',t'), \quad z = z(s',t')$$

where $s' = s'(x',y')$. At $t = 0$, this curve passes through both surfaces since it passes through the initial curve at the point

$$x(s',0) = x_0(s), \quad y(s',0) = y_0(s), \quad z(s',0) = z_0(s).$$

But if a characteristic curve has one point in common with an integral surface, it lies entirely on the surface. Thus, the characteristic curve lies on both surfaces, and in particular for t', we have finally

$$\phi'(x',y') = \phi'(x'(s',t'), y'(s',t')) = z(s',t') = \phi(s',t'). \qquad \square$$

When the initial curve is given by the intersection of two surfaces

$$\Phi_1(x,y,z) = 0, \quad \Phi(x,y,z) = 0, \tag{2.25}$$

we can obtained the solution using Eq. (2.20) and Eq. (2.25) to determine c_1, c_2 and the functionality Φ of Eq. (2.21). Note that the condition Eq. (2.22) implies the condition that the initial conditions must not be given on any characteristic curve.

Example 2.3 Find the integral surface of the equation

$$x\frac{\partial z}{\partial y} - y\frac{\partial z}{\partial x} = 0 \tag{2.26}$$

passing through the curve

$$x = 0, \quad y = z^2. \tag{2.27}$$

We have three ways to determine the Cauchy problem, one is using the parametric solutions, the other is using relation (2.21), and the last one is using Eq. (2.25) (which is equivalent to the latter one). Let us explore each of these ways.

i) Parametric solutions. The characteristic curves are given by equations

$$\frac{dx}{dt} = -y, \quad \frac{dy}{dt} = x, \quad \frac{dz}{dt} = 0 \tag{2.28}$$

whose solutions are

$$x = ae^{it} + be^{-it}, \quad y = i(ae^{it} - be^{-it}), \quad z = c_1, \tag{2.29}$$

where a, b, c are constants. The initial curve is given as

$$x_0(s) = 0, \quad y_0(s) = s, \quad z_0(s) = s^2. \tag{2.30}$$

Using these in Eq. (2.29), we obtain the family of characteristic

$$x(s,t) = s\sin t, \quad y(s,t) = s\cos t, \quad z(s,t) = s^2. \tag{2.31}$$

The Jacobian is

$$\frac{\partial(x,y)}{\partial(s,t)}\Big|_t = s$$

then for $s \neq 0$, we have

$$s^2 = x^2 + y^2,$$

and the integral surface, from Eq. (2.31), is given by

$$z(x,y) = x^2 + y^2.$$

ii) Using the functionality of one characteristic with respect the other.
The equations for the characteristic curves are

$$\frac{dx}{-y} = \frac{dy}{x} = \frac{dz}{0}$$

whose solutions are

$$x^2 + y^2 = c_1, \quad z = c_2. \tag{2.32}$$

According with Eq. (2.21)

$$z = c_2(c_1) = c_2(x^2 + y^2),$$

and using the initial conditions (2.27), we obtain

$$s^2 = c_2(s^2) \quad \text{or} \quad c_2(x^2 + y^2) = x^2 + y^2$$

then, the solution is

$$z(x, y) = x^2 + y^2.$$

iii) Using the system (2.25) and (2.20).
We have the following set the equations

$$x = 0, \quad z = y^2, \quad x^2 + y^2 = c_1 \quad \text{and} \quad z = c_2$$

where we obtain

$$c_1 = y^2 = z = c_2$$

then, the solution is again the same

$$z(x, y) = x^2 + y^2.$$

2.4   Quasi-Linear PDEFO for Functions Defined in $\Omega \subset \mathbb{R}^n$

A quasi-linear PDEFO defined in $\Omega \subset \mathbb{R}^n$ may be written as

$$\sum_{i=1}^{n} R_i(\mathbf{x}, z) \frac{\partial z}{\partial x_i} = Q(\mathbf{x}, z), \tag{2.33}$$

where $\mathbf{x} = (x_1, ..., x_n)$ is a point in $\Omega \subset \mathbb{R}^n$, $z = z(\mathbf{x})$ is a function defined in Ω with scalar values (in $\mathbb{R}$) which is at least once continuously differentiable. A solution $z = f(\mathbf{x})$ defines a n-dimensional integral surface in $\mathbb{R}^{n+1}$ whose normal direction,

$$\frac{(\nabla z, -1)}{\sqrt{1 + \nabla z \cdot \nabla z}},$$

$(\nabla z = (z_1, ..., z_n)$ with $z_i = \partial z/\partial x_i)$ is orthogonal to the vector field $\mathbf{E}$ defined by

$$\mathbf{E} = (R_1, ..., R_n, Q).$$

Thus, this field is tangent to the integral surface, and we may look for family of curves on $\mathbb{R}^{n+1}$ whose tangent direction, $d\vec{\xi} = (dx_1, ..., dx_n, dz)$ coincides with the direction of $\mathbf{E}$.

These curves are called characteristic curves and are given by equations

$$\frac{dx_1}{R_1} = ... = \frac{dx_n}{R_n} = \frac{dz}{Q} \tag{2.34a}$$

or the parametric equations

$$\frac{dx_i}{dt} = R_i(\mathbf{x}, z) \quad \text{for} \quad i = 1, ..., n \quad \frac{dz}{dt} = Q(\mathbf{x}, z), \tag{2.34b}$$

where "t" is the parameter which characterizes these curves. If initial conditions at $t = 0$ are characterized by the parameters $\mathbf{s}^* = (s_1, ..., s_{n-1}) \in \mathbb{R}^{n-1}$,

$$x_{i0}(\mathbf{s}^*) = x_i(0) \quad \text{for} \quad i = 1, ..., n \quad z_0(\mathbf{s}^*) = z(0), \tag{2.35}$$

the solution of (2.34b) would be expressed as

$$x_i = x_i(\mathbf{s}^*, t) \quad \text{for} \quad i = 1, ..., n \quad z = z(\mathbf{s}^*, t). \tag{2.36}$$

Thus, choosing the condition that the Jacobian of the transformation between $\mathbf{x}$ and $(\mathbf{s}^*, t)$ to be different from zero,

$$\left. \frac{\partial(x_1, ..., x_{n-1}, x_n)}{\partial(s_1, ..., s_{n-1}, t)} \right|_{t=0} \neq 0 \tag{2.37a}$$

(equivalent to the condition (2.22)), the inverse transformation,

$$\mathbf{s}^* = \mathbf{s}^*(\mathbf{x}), \quad t = t(\mathbf{x}), \tag{2.37b}$$

is obtained, and the solution of (2.33) with the conditions (2.35) is obtained as

$$\psi(\mathbf{x}) = z\left(\mathbf{s}^*(\mathbf{x}), t(\mathbf{x})\right). \tag{2.38}$$

The theorem of existence and uniqueness (theorem 2.1) is extended immediately to this case.

On the other hand, from (2.34a) one may obtain n-characteristics

$$\phi_i(\mathbf{x}) = C_i \quad \text{for} \quad i = 1, ..., n$$

and the general solution of (2.33) can be written as

$$\Phi\left(\phi_1(\mathbf{x}), ..., \phi_n(\mathbf{x})\right) = 0, \tag{2.39}$$

where the function Φ is arbitrary. Let us extend the theorem 2.1 to $\mathbb{R}^n$.

Theorem 2.2. *Consider the first order quasi-linear partial differential equation (2.33), where* R_i, Q, *have continuous partial derivatives with respect* x_i. *Suppose that along the initial hypersurface parametrized by* $\mathbf{s}^* \in \mathbb{R}^{n-1}$, $x_{0i} = x_{0i}(\mathbf{s}^*)$ *for* $i = 1, ..., n-1$, *and values* $z = z_0(\mathbf{s}^*)$ *are prescribed,* x_{0i}, z_0 *being continuously differentiable function on each* s_j, $j = 1, ..., n-1$. *Furthermore, let these function satisfy the expression Eq. (2.37a). Then, there exist one and only one solution* $z(\mathbf{x})$ *defined in some neighborhood of the initial hypersurface* Γ_o, *which satisfies the partial differential equation and the initial condition* $z(\mathbf{x}_0(\mathbf{s}^*)) = z_0(\mathbf{s}^*)$.

Proof. From the existence and uniqueness theorem for ordinary differential equations we may solve equations Eq. (2.34b) given the initial conditions Eq. (2.35). Their solutions Eq. (2.36) have derivatives with respect to the parameter s_j for $j = 1, ..., n-1$ and t which are continuous such that they satisfy the initial conditions

$$x_i(\mathbf{s}^*, 0) = x_{0i}(\mathbf{s}^*) \text{ , for } \quad i = 1, ..., n \quad \text{and} \quad z(\mathbf{s}^*, 0) = z_0(\mathbf{s}^*) \ .$$

Since one has the condition Eq. (2.37a) is satisfied, one can have the inverse relation Eq. (2.37b). Therefore, the function Eq. (2.38) is our candidate for solution of the problem. Indeed, one has (using Eq. (2.34a) and the linear independence of the variables s_j and t)

$$
\begin{aligned}
\sum_k R_k \frac{\partial \psi}{\partial x_k} &= \sum_{k=1}^{n} R_k \left[\sum_{l=1}^{n-1} \frac{\partial z}{\partial s_l} \frac{\partial s_l}{\partial x_k} + \frac{\partial z}{\partial t} \frac{\partial t}{\partial x_k} \right] \\
&= \sum_{k=1}^{n} \sum_{l=1}^{n-1} R_k \frac{\partial z}{\partial s_l} \frac{\partial s_l}{\partial x_k} + \frac{\partial z}{\partial t} \sum_{k=1}^{n} R_k \frac{\partial t}{\partial x_k} \\
&= \sum_{k=1}^{n} \sum_{l=1}^{n-1} \frac{\partial z}{\partial s_l} \frac{\partial x_k}{\partial t} \frac{\partial s_l}{\partial x_k} + \frac{\partial z}{\partial t} \sum_{k=1}^{n} \frac{\partial x_k}{\partial t} \frac{\partial t}{\partial x_k} \\
&= \sum_{l=1}^{n-1} \frac{\partial z}{\partial s_l} \left(\sum_{k=1}^{n} \frac{\partial x_k}{\partial t} \frac{\partial s_l}{\partial x_k} \right) + \frac{\partial z}{\partial t} \frac{dt}{dt} \\
&= \sum_{l=1}^{n-1} \frac{\partial z}{\partial s_l} \left(\frac{ds_l}{dt} \right) + \frac{\partial z}{\partial t} \\
&= Q \ .
\end{aligned}
$$

Therefore, the function Eq. (2.38) is a solution of the equation Eq. (2.33). Now, this solution is unique since assume there is another one $\psi'(\mathbf{x})$ satisfying Eq. (2.33) which pass through the hypersurface $\Gamma_o(\mathbf{s}^*)$. Then, since one must have that $\psi'(\mathbf{x}')|_{t=0} = \psi(\mathbf{x})|_{t=0}$ for any $\mathbf{x}', \mathbf{x} \in \mathbb{R}^n$ and the solutions are defined along the whole characteristics, one must have that $\psi'(\mathbf{x}) = \psi(\mathbf{x})$ for any $\mathbf{x} \in \mathbb{R}^n$. $\qquad\qquad\qquad\qquad\square$

Example 2.4 Find the integral surface of the equation

$$\sum_{i=1}^{n} x_i \frac{\partial z}{\partial x_i} = \alpha z, \qquad (2.40)$$

where $\alpha \in \mathbb{R}$ is given. The equations for the characteristic are

$$\frac{dx_1}{x_1} = \ldots = \frac{dx_n}{x_n} = \frac{dz}{\alpha z}. \qquad (2.41)$$

Taking the nth-term together with each previous one (this selection is arbitrary), one gets the $n-1$ characteristics

$$C_i = \frac{x_i}{x_n} \quad \text{for} \quad i = 1, ..., n-1.$$

Taking the last two terms of (2.41), one gets the characteristic

$$C_n = \frac{z}{x_n^\alpha}.$$

Thus, the general solution of (2.40) is given by

$$\Phi\left(\frac{x_1}{x_n}, \frac{x_2}{x_n}, ..., \frac{x_{n-1}}{x_n}, \frac{z}{x_n} \right) = 0.$$

This solution can also be written as

$$z = x_n^\alpha \Psi\left(\frac{x_1}{x_n}, \frac{x_2}{x_n}, ..., \frac{x_{n-1}}{x_n} \right) \qquad (2.42)$$

This function is an homogeneous function of order α since for any $\lambda \in \mathbb{R}$, it follows that

$$z(\lambda \mathbf{x}) = \lambda^\alpha z(\mathbf{x}).$$

2.5   Problems

2.1   Make the classification of the following partial differential equations

i) $(x_1 - x_2)^{1/2} \left(\dfrac{\partial^2 z}{\partial x_i^2} \right)^2 = 0, \quad z = z(x_1, x_2)$

ii) $x_1 e^z \dfrac{\partial z}{\partial x_2} + x_2 \dfrac{\partial z}{\partial x_1} + x_1^2 x_2 z = 0 \quad z = z(x_1, x_2)$

iii) $(x_1^2 + x_2^2 + x_3^2) \dfrac{\partial^2 z}{\partial x_1^2} + \dfrac{\partial z}{\partial x_2} + b(x_1) \dfrac{\partial z}{\partial x_3} = 0, \quad z = z(x_1, x_2, x_3)$

iv) $\dfrac{\partial u}{\partial t} + u \dfrac{\partial u}{\partial x} = 0 \quad u = u(x, t)$

v) $x_1 \dfrac{\partial z}{\partial x_2} - x_2 \dfrac{\partial z}{\partial x_1} = mz \quad z = z(x_1, x_2)$.

2.2   Demonstrate that the following PDEFO are linear

i) $y \dfrac{\partial L}{\partial x} - y \dfrac{\partial L}{\partial y} = 0 , \quad L = L(x, y)$

ii) $x^2 \dfrac{\partial u}{\partial x} + y^2 \dfrac{\partial u}{\partial y} + z^2 \dfrac{\partial u}{\partial z} = \pi z , \quad u = u(x, y, z)$

iii) $v \dfrac{\partial K}{\partial x} + F(x, v, t) \dfrac{\partial K}{\partial v} + \dfrac{\partial K}{\partial t} = 0 , \quad K = K(x, v, t)$.

2.3   Prove that the function ϕ is solution of its respective PDEFO

i) $\phi(x, y) = \dfrac{1}{2}(x^2 + y^2) ; \quad x \dfrac{\partial z}{\partial y} - y \dfrac{\partial z}{\partial x} = 0$

ii) $\phi(x, y) = \dfrac{1}{2}(x^2 - y^2) , \quad x \dfrac{\partial z}{\partial y} + y \dfrac{\partial z}{\partial x} = 0$

iii) $\phi(x, y, v_x, v_y) = A_1 \left(x, y, \dfrac{v_y}{v_x}, v_x \right) v_x + A_2 \left(x, y, v_y, \dfrac{v_x}{v_y} \right) v_y$

$v_x \dfrac{\partial L}{\partial v_x} + v_y \dfrac{\partial L}{\partial v_y} + L = 0 , \quad L = L(x, y, v_x, v_y)$.

2.4   Find the integral surface of the following PDEFO:

i) $x \dfrac{\partial z}{\partial x} + y \dfrac{\partial z}{\partial y} = \alpha z , \quad z = z(x, y)$

ii) $y^2 \dfrac{\partial z}{\partial x} - x^3 \dfrac{\partial z}{\partial y} = 0 , \quad z = z(x, y)$

iii) $\dfrac{\partial u}{\partial t} + \dfrac{\partial u}{\partial x} = 0 , \quad u = u(x, t)$

iv) $y^\alpha \dfrac{\partial z}{\partial x} + x^{\alpha+1} \dfrac{\partial z}{\partial y} = \alpha z , \quad z = z(x, y), \quad \alpha \geq 1$.

2.5 Find the integral surfaces of the following PDEFO which pass through the respective curve Γ_o, and determine whether or not Γ_o is a characteristic curve of the equation:

i) $\quad x\dfrac{\partial z}{\partial x} + y\dfrac{\partial z}{\partial y} = \alpha z$, $\quad \Gamma_o : \{y = x, z = x/2\}$

ii) $\quad \dfrac{\partial L}{\partial v} - L = K(x,v)$, $\quad \Gamma_o : \{x = v, L = 0\}$

iii) $\quad x\dfrac{\partial z}{\partial y} - y\dfrac{\partial z}{\partial x} = \alpha z$, $\quad \Gamma_o\{x^2 + y^2 = 4, z = x^2\}$

iv) $\quad \dfrac{1}{c}\dfrac{\partial u}{\partial t} + \dfrac{\partial u}{\partial x} = f(x)u$, $\quad \Gamma_o : \{t = 0, u = 0\}$

v) $\quad x^{1/3}\dfrac{\partial z}{\partial y} - \dfrac{\partial z}{\partial x} = 0$, $\quad \Gamma_o(s) : \{x_o(s) = 1, y_o(s) = s, z_o(s) = \sqrt{1 - s^2}\}$.

2.6 Giving the curve of initial data, Γ_o, show that the following problems have not solutions:

i) $\quad \dfrac{\partial z}{\partial x} - \dfrac{1}{y}\dfrac{\partial z}{\partial y} = 0$, $\quad \Gamma_o : \{x + y^2 = 1, z = x^2\}$

ii) $\quad v\dfrac{\partial L}{\partial v} - L = K(x,v)$, $\quad \Gamma_o : \{x = 0, L = v^2/2\}$

iii) $\quad x^2\dfrac{\partial z}{\partial y} - y^2\dfrac{\partial z}{\partial x} = x^2 z$, $\quad \Gamma_o : \{y = (1 - x^3)^{1/3}, z = x^2 e^{-x}\}$

iv) $\quad v\dfrac{\partial K}{\partial x} - x\dfrac{\partial K}{\partial v} = 0$, $\quad \Gamma_o : \{x^2 + v^2 = 1/3, K = v^2/2\}$.

2.7 Demonstrate that the solution of the PDEFO defined in $\mathbb{R}^n$ by

$$\sum_{i=1}^{n} x_i \frac{\partial z}{\partial x_i} = -\alpha z , \qquad \alpha \geq 0$$

such that pass through the hyperplane $\Gamma_o : \{x_1 = 1\}$, having the value there as $z = \phi(x_2, \ldots, x_n)$, with ϕ being any differentiable function, is given by

$$z(\mathbf{x}) = \phi(x_1, \ldots, x_n)e^{\alpha(1 - x_1)} .$$

2.8 Demonstrate that the solution of the PDEFO $z^2\,(\partial z/\partial x) - \partial z/\partial y = z$, such that at $y = 0$, $z(x, 0) = \phi(x)$ is an invertible function in the domain $\Omega \subset \mathbb{R}^2$ defined by the equation, is solution of the following transcendental algebraic equation

$$z^2(1 - e^{2y}) + 2\phi^{-1}(ze^y) - 2x = 0$$

2.9 Demonstrate that the solution of the PDEFO $u_t - u u_x = 0$ defined in $\mathbb{R}^2$, $u = u(x,t)$, such that $u(x,0) = \phi(x)$, has the implicit solution

$$u(x,t) = \phi\big(x - u(x,t)t\big) \ .$$

In addition, suppose that $\phi(x)$ is of the form

$$\phi(x) = \frac{1}{\sigma\sqrt{2\pi}} e^{-x^2/2\sigma^2} \ .$$

Then, make a plot of the solution ($\sigma = 1$).

2.10 Find the solution of the PDEFO defined in $\mathbb{R}^3$ by

$$x\frac{\partial u}{\partial y} - y\frac{\partial u}{\partial y} + u\frac{\partial u}{\partial z} = 0 \ ,$$

with $u = u(x,y,z)$ and such that $u(0,y,z) = \psi(y,z)$, and show that this solution is given implicitly as

$$u(x,y,z) = \phi\left(\sqrt{x^2 + y^2}, \pm u(x,y,z) \arcsin \frac{x}{\sqrt{x^2+y^2}} + z \right) \ .$$

2.11 Find the solution in parametric form of the PDEFO defined in $\mathbb{R}^3$ by

$$x_1\frac{\partial z}{\partial x_1} + x_2\frac{\partial z}{\partial x_2} + x_3\frac{\partial z}{\partial x_3} = z^2$$

which pass through the initial surface defined as $\Gamma_o(\mathbf{s}^*) : \{x_{1o}(\mathbf{s}^*) = 1, x_{2o}(\mathbf{s}^*) = s_1, x_{3o}(\mathbf{s}^*) = s_2, z_o(\mathbf{s}^*) = s_1 + s_2\}$, and show that the solution can be written as

$$z(\mathbf{x}) = \frac{x_2 + x_3}{x_1 - (x_2 + x_3)\ln x_1} \ .$$

2.12 Find the solution of the quasi-linear PDFEO

$$u\frac{\partial u}{\partial x} + y\frac{\partial u}{\partial y} + z\frac{\partial u}{\partial z} = u^2$$

such that it passes through the initial surface $\Gamma_o : \{x = 0, u = y + z\}$.

2.6 *References*

2.1 F. John, *Partial Differential Equations*, Springer-Verlag, 1978. Chap.1.

2.2 L. Elsgoltz, *Ecuaciones Diferenciales y Cálculo Variacional*, Ediciones Cultura Popular, 1975. Pages 248-260.

2.3 R. Courant and D. Hilbert, *Methods of Mathematical Physics*, John Wiley & Sons, 1962. Vol. II chapters I and II.

Chapter 3

Physical Applications I

In this chapter we illustrate the uses of the linear partial differential equations of first order in several topics of Physics. The purpose of this chapter is to motivate the importance of this branch of mathematics into the physical sciences. In most of the applications, it is not intended to fully develop the consequences and the theory involved in the applications, but usually we point out some reference where the physical theory can be seen.

3.1 Mechanics

As we saw in chapter 1.6, there is a connection between Newton's equations of motion (1.58), Euler's equation (1.62), the Lagrangian (1.61) and the Hamiltonian (1.64) of the system. One of the most important objectives of Mechanics is not only to give the solution of the equation of motion in either form (1.58), (1.65) or (1.76), but also to find the constant of motion of the system. In terms of Eq. (1.58) and for time independent forces (autonomous systems), we have

$$\frac{dv_i}{dt} = f_i(\mathbf{q}, \mathbf{v}), \quad i = 1, ..., N \tag{3.1a}$$

and

$$\frac{dq_i}{dt} = v_i, \quad i = 1, ..., N. \tag{3.1b}$$

By a constant of motion we mean any surface $K = K(\mathbf{q}, \mathbf{v})$ defined in $2N$-dimensional space $(\mathbf{q}, \mathbf{v})$ such that is normal direction projection on this space is the perpendicular to the vector field (1.59b),

$$\mathbf{E} = (\mathbf{v}, \mathbf{f}), \tag{3.2}$$

where $\mathbf{v} = (v_1, ..., v_N)$ and $\mathbf{f} = (f_1, ..., f_N)$. That is, the function K satisfies the linear partial differential equation

$$\sum_{i=1}^{N} \left(v_i \frac{\partial K}{\partial q_i} + f_i(\mathbf{q}, \mathbf{v}) \frac{\partial K}{\partial v_i} \right) = 0, \tag{3.3}$$

and the total time derivative operator along $\mathbf{E}$ is defined as

$$\frac{d}{dt} = \sum_{i=1}^{N} \left(v_i \frac{\partial}{\partial q_i} + f_i(\mathbf{q}, \mathbf{v}) \frac{\partial}{\partial v_i} \right). \tag{3.4}$$

As we saw in chapter 2, the solution of Eq. (3.3) is given by

$$K = K(c_1, ..., c_{2N-1}), \tag{3.5}$$

where $c_i (i = 1, ..., 2N - 1)$ are the characteristic curves, which are solutions of equations

$$\frac{dq_1}{v_1} = ... = \frac{dq_N}{v_N} = \frac{dv_1}{f_1} = ... = \frac{dv_N}{f_N}. \tag{3.6}$$

Example 3.1. Find the constant of motion of one dimensional single particle of mass "m" moving in a dissipative medium whose frictional force is proportional to the velocity squared.

The equation of motion can be written as the following dynamical system, Eq. (1.84),

$$\frac{dv}{dt} = -\frac{\alpha}{m} v^2$$

and

$$\frac{dq}{dt} = v,$$

where α is the friction constant and m is the mass os the particle, and $v > 0$. The equation of the characteristic curve is given by

$$\frac{dq}{v} = \frac{dv}{-(\alpha/m)v^2}$$

whose solution gives us the characteristic curves

$$c = v e^{\alpha/mq}. \tag{3.7}$$

Thus, the constant of motion is given by an arbitrary function of "c",

$$K = K(v e^{\alpha/mq}). \tag{3.8}$$

As we can see from Eq. (3.5) or Eq. (3.8), there exists a non-numerable set of constants of motion for a given physical system. One possible way to select a proper function is using the criteria of reducibility. It consists in choosing the functionality K in such a way that we can get the known form of the energy ($E = mv^2/2 + V(q)$) or some other constant of motion when the parameters, which characterize the nonconservative force, go to zero. In the above example if we take

$$K = \frac{1}{2}mc^2 = \frac{1}{2}mv^2 e^{2\alpha q/m}, \tag{3.9}$$

we obtain the usual energy of a free particle for α going to zero. Thus, one can use this constant of motion as the constant of motion which contains the proper dissipative dynamics of the system.

If the Hamiltonian in the expression (1.58) is substituted by the constant of motion (3.5) and relation (1.63) is used. Then, given the constant of motion, we obtain the following linear partial differential equation for the Lagrangian

$$K(\mathbf{q}, \mathbf{v}) = \sum_{i=1}^{N} \dot{q}\frac{\partial L}{\partial \dot{q}} - L. \tag{3.10}$$

The equations for the characteristic are

$$\frac{dq_1}{0} = ... = \frac{dq_N}{0} = \frac{d\dot{q}_1}{\dot{q}_1} = ... = \frac{d\dot{q}_N}{\dot{q}_N} = \frac{dL}{L + K}. \tag{3.11}$$

From the first $2N$ terms, we obtain

$$C_i = q_i \quad i = 1, ..., N \tag{3.12a}$$

and

$$C_{ij} = \dot{q}_i/\dot{q}_j \tag{3.12b}$$

such that

$$C_{ji}C_{ij} = 1 \quad i \neq j = 1, ..., N. \tag{3.13}$$

Now, making use of Eq. (3.12) and Eq. (3.13) in the last term of Eq. (3.11) and integrating with any of the N-terms of its left hand side, we get

$$L = \sum_{i=1}^{N} A_i(\mathbf{C}, \mathbf{C}^{(i)})\dot{q}_i$$

$$+ \frac{1}{N}\sum_{i=1}^{N} \dot{q}_i \int^{\dot{q}_i} \frac{K(\mathbf{C}, C_{1i}, C_{2i}, ..., \xi, C_{i+1,i}, ..., C_{iN})}{\xi^2} d\xi \tag{3.14}$$

where A_i are arbitrary functions of the characteristics $\mathbf{C} = (C_1, ..., C_N)$ and $\mathbf{C}^{(i)} = (C_{1i}, C_{2i}, ..., C_{i-1,i}, C_{i+1,i}, ..., C_{Ni})$. The first on the right hand side of Eq. (3.14) is the solution of the homogeneous part of Eq. (3.10), that is when K is equal to zero. This can be seen using the fact that

$$\sum_{i=1}^{N} \dot{q}_i \frac{\partial A_i}{\partial \dot{q}_i} = 0. \tag{3.15}$$

And the second term on the right hand side of Eq. (3.14) represents the solution of the inhomogeneous equation (3.10). The factor N appearing in this term is necessary to accomplish this. The generalized linear momentum (1.63) is

$$p_j = A_j(\mathbf{C}, \mathbf{C}^{(i)}) + \frac{1}{N} \int^{\dot{q}_j} \frac{K(\mathbf{C}, C_{1j}, ..., C_{j-1,j}, \xi, C_{j+1,j}, ..., C_{jN})}{\xi^2} d\xi + \frac{K}{N\dot{q}_j}$$

and after making an integration by parts, we get

$$p_j = A_j + \frac{1}{N} \int^{\dot{q}_j} \frac{\partial K(\mathbf{C}, C_{1j}, ..., C_{j-1,j}, \xi, C_{j+1,j}, ..., C_{jN})}{\partial \xi} \frac{d\xi}{\xi^2}. \tag{3.16}$$

Example 3.2. Find the Lagrangian and the generalized linear momentum for the system described in the Example 3.1 with the constant of motion given by Eq. (3.9).

According to Eq. (3.14), the Lagrangian for one-dimensional autonomous systems can be calculated through

$$L(q, v) = A(q)v + v \int^{v} \frac{K(q, \xi)}{\xi^2} d\xi, \tag{3.17a}$$

if

$$K(q, v) = \frac{1}{2} m v^2 e^{2\alpha q/m} \tag{3.17b}$$

is the constant of motion, we have the Lagrangian and the general momentum as

$$L = A(q, v)v + \frac{1}{2} m v^2 e^{2\alpha q/m} \tag{3.18}$$

and

$$p = A(q, v) + m v e^{2\alpha q/m}. \tag{3.19}$$

The first term of the right hand side of Eq. (3.17a) is the total time derivative of a function depending on the variable q. As it was pointed out for Eq. (1.72), this term gives us an equivalent Lagrangian therefore, it can be neglected.

Let us now apply the operator (3.4) to expression (3.10) to look for a relation between expression (3.10) and Euler's equation (1.62),

$$\frac{dK}{dt} = \sum_{i=1}^{N}\sum_{n=1}^{N}\left[\dot{q}_i\dot{q}_n\frac{\partial^2 L}{\partial\dot{q}_i\partial q_n} + f_n\dot{q}_i\frac{\partial^2 L}{\partial\dot{q}_i\partial\dot{q}_n}\right] - \sum_{i=1}^{N}\dot{q}_i\frac{\partial L}{\partial\dot{q}_i},$$

then after rearranging terms

$$\frac{dK}{dt} = \sum_{i=1}^{N}\dot{q}_i\left(\sum_{n=1}^{N}\left[\dot{q}_n\frac{\partial}{\partial q_n}\left(\frac{\partial L}{\partial\dot{q}_i}\right) + f_n\frac{\partial}{\partial\dot{q}_n}\left(\frac{\partial L}{\partial\dot{q}_i}\right)\right] - \frac{\partial L}{\partial\dot{q}_i}\right)$$

and according to definition (3.14), the above expression can be written as

$$\frac{dK}{dt} = \sum_{i=1}^{N}\dot{q}_i\left[\frac{d}{dt}\left(\frac{\partial L}{\partial\dot{q}_i}\right) - \frac{\partial L}{\partial q_i}\right]. \tag{3.20}$$

Let $\mathbf{V}$ and $\mathbf{E}$ be the vectors defined as

$$\mathbf{V} = (\dot{q}_1, ..., \dot{q}_n)$$

and

$$\mathbf{E} = \left(\frac{d}{dt}\left(\frac{\partial L}{\partial\dot{q}_1}\right) - \frac{\partial L}{\partial q_1}, ..., \frac{d}{dt}\left(\frac{\partial L}{\partial\dot{q}_n}\right) - \frac{\partial L}{\partial q_n}\right),$$

then, Eq. (3.20) expresses the fact that whenever K is a constant of motion of the system, the orthogonality relation

$$\mathbf{V}\cdot\mathbf{E} = 0 \tag{3.21}$$

is always satisfied. We must notice that this does not guarantee the existence of the Lagrangian except for the one-dimensional case. For $\dot{q} \neq 0$, K in a one-dimensional case, K is a constant of the motion if and only if L satisfies the Euler's equation (1.62). This means that we can always find the Lagrangian by using Eq. (3.17). For the n-dimensional case, we can use expression (3.14) and Euler's equation to find restrictions on the functions A_i and the constant of motion K in order for the L function to represent the Lagrangian of the system (3.1). Thus, developing the term

$$\sum_{n=1}^{N}\left(\dot{q}_n\frac{\partial}{\partial q_n} + f_n\frac{\partial}{\partial\dot{q}_n}\right)\left(\frac{\partial L}{\partial\dot{q}_l}\right) - \frac{\partial L}{\partial q_l} = 0$$

with L given by the expression (3.14), we have

$$\sum_{n=1}^{N}\left[\dot{q}_n\frac{\partial A_l}{\partial q_n}+f_n\frac{\partial A_l}{\partial \dot{q}_n}\right]-\sum_{i=1}^{N}\frac{\partial A_i}{\partial q_l}\dot{q}_i+\frac{1}{N}\sum_{n=1}^{N}\dot{q}_n\int^{\dot{q}_l}\frac{\partial K}{\partial q_n}\frac{d\xi}{\xi^2}$$

$$-\frac{1}{N}\sum_{i=1}^{N}\dot{q}_i\int^{\dot{q}_i}\frac{\partial K}{\partial q_l}\frac{d\xi}{\xi^2}+\frac{1}{N\dot{q}_l}\sum_{n=1}^{N}\left[\dot{q}_n\frac{\partial A_l}{\partial q_n}+f_n\frac{\partial A_l}{\partial \dot{q}_n}\right]=0,$$

and using the fact that K is a constant of motion, being A_i an arbitrary independent functions of K, we get

$$\sum_{n=1}^{N}\dot{q}_n\left[\int^{\dot{q}_l}\frac{\partial K}{\partial q_n}\frac{d\xi}{\xi^2}-\int^{\dot{q}_n}\frac{\partial K}{\partial q_l}\frac{d\xi}{\xi^2}\right]=0 \qquad (3.22a)$$

and

$$\sum_{n=1}^{N}\dot{q}_n\left[\frac{\partial A_l}{\partial q_n}+\frac{\partial A_n}{\partial \dot{q}_l}\right]+\sum_{n=1}^{N}f_n\frac{\partial A_l}{\partial \dot{q}_n}=0 \qquad (3.22b)$$

for every $l=1,...,N$. So, if the conditions (3.22) are satisfied, the function (3.14) represents a Lagrangian for the system (3.1).

Example 3.3. Find a constant of the motion for a particle moving in three dimensional space under a field of forces derivable from a potential function $\phi = \phi(\mathbf{q})$. Verify the relation (3.22) and find a Lagrangian.

The equations of motion (3.1) are

$$\frac{dv_i}{dt}=-\frac{1}{m}\frac{\partial \phi}{\partial q_i} \quad i=1,2,3 \qquad (3.23a)$$

and

$$\frac{dq_i}{dt}=v_i \quad i=1,2,3 \qquad (3.23b)$$

where m is the mass of the particle. The equations for the characteristic curves of the constant of motion (3.6) are

$$\frac{dq_1}{v_1}=\frac{dq_2}{v_2}=\frac{dq_3}{v_3}=\frac{mdv_1}{-(\partial\phi/\partial q_1)}=\frac{mdv_2}{-(\partial\phi/\partial q_2)}=\frac{mdv_3}{-(\partial\phi/\partial q_3)}$$

which can be completed to yield the exact differential

$$d\left(\phi+\frac{1}{2}m(v_1^2+v_2^2+v_3^2)\right)=0.$$

Then, a possible constant of motion is

$$K=\frac{1}{2}m(v_1^2+v_2^2+v_3^2)+\phi. \qquad (3.24)$$

From Eq. (3.23a), we see that

$$3v_l\left(\frac{\partial\phi}{\partial q_l}\right) = -3\frac{d}{dt}\left(\frac{1}{2}mv_l^2\right) = -\frac{d}{dt}\left(\frac{1}{2}m(v_1^2 + v_2^2 + v_3^2)\right), \qquad (3.25)$$

but taking the time derivative of (3.24), we obtain

$$\frac{d}{dt}\left[\frac{1}{2}m(v_1^2 + v_2^2 + v_3^2)\right] = -\frac{d\phi}{dt} = -\sum_{n=1}^{3}\left(\frac{\partial\phi}{\partial q_l}\right)\dot{q}_n$$

so we obtain for Eq. (3.25)

$$3\dot{q}_l\left(\frac{\partial\phi}{\partial q_l}\right) = \sum_{n=1}^{3}\left(\frac{\partial\phi}{\partial q_n}\right)\dot{q}_n \quad \text{for} \quad l = 1, 2, 3. \qquad (3.26)$$

Using Eq. (3.25) in Eq. (3.22a), we have

$$\sum_{n=1}^{3}\dot{q}_n\left[-\left(\frac{\partial\phi}{\partial q_n}\right)\frac{1}{\dot{q}_l} + \left(\frac{\partial\phi}{\partial q_l}\right)\frac{1}{\dot{q}_n}\right] = \frac{1}{\dot{q}_l}\left[-\sum_{n=1}^{3}\dot{q}_n\left(\frac{\partial\phi}{\partial q_n}\right) + 3\dot{q}_l\left(\frac{\partial\phi}{\partial\dot{q}_l}\right)\right]$$

which according to Eq. (3.26) is zero. Choosing the functions $A_i = 0$ ($i = 1, 2, 3$), the condition Eq. (3.22b) is also satisfied. Using now Eq. (3.24) in the expression (3.14), we get

$$L = \frac{1}{2}m(v_1^2 + v_2^2 + v_3^2) - \phi. \qquad (3.27)$$

Now, using the calculated generalized linear momentum (3.16), if we can make

$$\dot{q}_i = \dot{q}_i(\mathbf{q}, \mathbf{p}) \quad i = 1, ..., N \qquad (3.28)$$

then, we could use these functions in Eq. (3.5) to obtain the Hamiltonian associated with the system (3.1) as

$$H(\mathbf{q}, \mathbf{p}) = K(c_1(\mathbf{q}, \dot{\mathbf{q}}(\mathbf{q}, \mathbf{p})), ..., c_{2n-1}(\mathbf{q}, \dot{\mathbf{q}}(\mathbf{q}, \mathbf{p}))). \qquad (3.29)$$

Example 3.4. Calculate the Hamiltonian for the Example 3.2 and Example 3.2.

In the Example 3.2, we take $A(q) = 0$ in expression (3.19) to obtain

$$p = m\dot{q}e^{2\alpha q/m}.$$

Then, the inverse relation is

$$\dot{q} = \frac{p}{m}e^{-2\alpha q/m},$$

and substituting this expression in Eq. (3.17b), the Hamiltonian becomes

$$H = \frac{p^2}{2m}e^{-2\alpha q/m}.$$

In Example 3.3, we get from Eq. (3.27) and Eq. (1.63)

$$p_i = m\dot{q}_i \quad i = 1, 2, 3.$$

Substituting this in Eq. (3.24), we obtain the Hamiltonian

$$H = \frac{1}{2m}(p_1^2 + p_2^2 + p_3^2) + \phi.$$

3.2   Angular Momentum in Quantum Mechanics

In classical mechanics, the angular momentum vector of a particle with respect to a reference system S is defined as

$$\mathbf{L} = \mathbf{r} \times \mathbf{p} \tag{3.30}$$

where $\mathbf{r}$ is the position vector of the particle with respect to S and $\mathbf{p}$ is its linear momentum. In quantum mechanics, the components of the linear momentum are represented by the linear operators

$$p_x = -i\hbar \frac{\partial}{\partial x}, \quad p_y = -i\hbar \frac{\partial}{\partial y}, \quad \text{and} \quad p_z = -i\hbar \frac{\partial}{\partial z}, \tag{3.31}$$

where $\hbar$ is Planck's constant divided by 2π. Hence, the operator associated to the components of the angular momentum vector are given in cartesian coordinates (x, y, z) as

$$L_x = -i\hbar \left(y\frac{\partial}{\partial z} - z\frac{\partial}{\partial y} \right), \tag{3.32a}$$

$$L_y = -i\hbar \left(z\frac{\partial}{\partial x} - x\frac{\partial}{\partial z} \right), \tag{3.32b}$$

$$L_z = -i\hbar \left(x\frac{\partial}{\partial y} - y\frac{\partial}{\partial x} \right), \tag{3.32c}$$

and in spherical coordinates (r, θ, ϕ) they are given as follows

$$L_x = i\hbar \left(\sin\phi \frac{\partial}{\partial \theta} + \cot\theta \cos\phi \frac{\partial}{\partial \phi} \right), \tag{3.33a}$$

$$L_y = i\hbar \left(-\cos\phi \frac{\partial}{\partial \theta} + \cot\theta \sin\phi \frac{\partial}{\partial \phi} \right), \tag{3.33b}$$

$$L_z = -i\hbar \frac{\partial}{\partial \phi}, \tag{3.33c}$$

where the relation between cartesian and spherical coordinates is as following

$$x = r\sin\theta \cos\phi \tag{3.34a}$$
$$y = r\sin\theta \sin\phi \tag{3.34b}$$
$$y = r\cos\theta. \tag{3.34c}$$

The square of the total angular momentum is defines as

$$L^2 = L_x^2 + L_y^2 + L_z^2 \tag{3.35}$$

and is expressed in terms of spherical coordinates as

$$L^2 = -\hbar^2 \left[\frac{1}{\sin\theta} \frac{\partial}{\partial\theta} \left(\sin\theta \frac{\partial}{\partial\theta} \right) + \frac{1}{\sin^2\theta} \frac{\partial^2}{\partial\theta^2} \right]. \tag{3.36}$$

The operators (3.32) or (3.33) satisfy the following commutation rules

$$[L_i, L_i] = 0 \quad i = x, y, z, \tag{3.37a}$$

and

$$[L_x, L_y] = i\hbar L_z, \quad [L_y, L_z] = i\hbar L_x, \quad [L_z, L_x] = i\hbar L_y \tag{3.37b}$$

where the commutation between two operators A and B is a bilinear operation defined as

$$[A, B] = AB - BA \tag{3.38}$$

and it has the property

$$[AB, C] = A[B, C] + [A, C]B \tag{3.39}$$

for every A, B, C operators. Let us calculate $[L^2, L_x]$. Using the relations (3.35) and (3.37b)

$$[L^2, L_x] = [L_x^2, L_x] + [L_y^2, L_x] + [L_z^2, L_x]$$
$$= -i\hbar(L_y L_z + L_z L_y) + i\hbar(L_z L_y + L_y L_z) = 0$$

and in the same way

$$[L^2, L_y] = [L^2, L_z] = 0. \tag{3.40}$$

The relation(3.37) tell us that the set L_x, L_y, L_z from the Lie algebra associated to the group of rotations in three dimensional space, called $O(3)$, or also the Lie algebra of 2×2 complex matrix with determinant equal to one (so called $SU(2)$). This group leaves invariant the usual Euclidean metric in this space $(ds^2 = dx^2 + dy^2 + dz^2)$. Defining the new operators

$$L_+ = L_x + iL_y \tag{3.41a}$$

and

$$L_- = L_x - iL_y, \tag{3.41b}$$

the relations (3.37) and (3.40) take the standard form

$$[L_z, L_+] = \hbar L_+, \quad [L_z, L_-] = -\hbar L_-, \quad [L_+, L_-] = 2\hbar L_z \tag{3.42}$$

and

$$[L^2, L_+] = [L^2, L_-] = [L^2, L_z] = 0. \tag{3.43}$$

In spherical coordinates and using Eq. (3.33), L_+ and L_- are written as

$$L_+ = \hbar e^{i\phi}\left(\frac{\partial}{\partial\theta} + i\cot\theta\frac{\partial}{\partial\phi}\right) \tag{3.44}$$

and

$$L_- = \hbar e^{-i\phi}\left(-\frac{\partial}{\partial\theta} + i\cot\theta\frac{\partial}{\partial\phi}\right). \tag{3.45}$$

Let us choose the projection L_z of the angular momentum vector to do the quantization. Because of relation(3.43), we know that the operator L^2 and L_i have common eigenfunctions, but different eigenvalues, that is

$$L^2 U_{\lambda m} = \hbar^2 \lambda U_{\lambda m} \tag{3.46}$$

and

$$L_z U_{\lambda m} = \hbar m U_{\lambda m}, \tag{3.47}$$

with the normalization condition

$$\int U_{\lambda m}^\dagger U_{\lambda m} d\Omega = 1, \tag{3.48}$$

where $d\Omega = \sin\theta d\theta d\phi$ is the differential of the solid angle, $U_\lambda^\dagger$ represents the transpose conjugated operation with the properties: i) $f^\dagger = f^*$, ii) $(A + \alpha B)^\dagger = A^\dagger + \alpha^* B^\dagger$, and iii) $(AB)^\dagger = B^\dagger A^\dagger$ for any linear operators A and B, any complex number α, and ny complex function f. This expression is usually written as $\langle \lambda m | \lambda m \rangle = 1$. Using Eq. (3.33c), the solutions of Eq. (3.47) in spherical coordinates is

$$U_{\lambda m} = f_\lambda(\theta)e^{im\phi},$$

and in cartesian coordinates, using Eq. (3.32c), we have to solve the following linear partial differential equation

$$x\frac{\partial U_{\lambda m}}{\partial y} - y\frac{\partial U_{\lambda m}}{\partial x} = im U_{\lambda m}$$

from which the equations for the characteristic curves are

$$\frac{dy}{x} = \frac{dx}{-y} = \frac{dU_{\lambda m}}{im U_{\lambda m}}.$$

From the first two terms, we obtain the characteristic

$$c = x^2 + y^2,$$

and using this expression in the third term, the resulting equation is

$$im\int\frac{dy}{(c - y^2)^{1/2}} = \int\frac{dU_{\lambda m}}{U_{\lambda m}}$$

which has the solution

$$U_{\lambda m} = A_\lambda(x^2 + y^2) \exp\left[im \arcsin \frac{y}{(x^2+y^2)^{1/2}}\right].$$

If $m \in N$, the solution is periodic with respect the angle ϕ. To find the complete solution and the eigenvalues λ of the operator L^2, let us use the algebraic relations (3.42) The action of the operators L_+ and L_- on the function $U_{\lambda m}$ can be know through the action $L_z L_\pm U_{\lambda m}$ and using Eq. (3.42) and Eq. (3.47). The operator (3.32) and (3.36) are Hermitian. One say that an operator A is Hermitian if it satisfies the relation

$$\int U_{\lambda m}^\dagger (AU_{\lambda m})d\Omega = \int (AU_{\lambda m})^\dagger U_{\lambda m} d\Omega, \qquad (3.49a)$$

that is $A^\dagger = A$. Note, however, that the operators L_+ and L_- are not Hermitian, in fact

$$L_+^\dagger = L_- \qquad (3.49b)$$

and

$$L_-^\dagger = L_+. \qquad (3.49c)$$

Adding and subtracting the operator $L_\pm L_z$ and using Eq. (3.42), if follows

$$L_z L_\pm U_{\lambda m} = ([L_z, L_\pm] + L_\pm L_z)U_{\lambda m} = (m \pm 1)\hbar(L_\pm U_{\lambda m}),$$

in other words, the function $L_\pm U_{\lambda m}$ is also an eigenvector of the operator L_z, but with eigenvalues $m \pm 1$,

$$L_z(L_\pm U_{\lambda m}) = \hbar(m \pm 1)(L_\pm U_{\lambda m}),$$

this means that the function $L_\pm U_{\lambda m}$ is proportional to the function $U_{\lambda m \pm 1}$. if $a_{\lambda m}^\pm$ is the constant of proportionality, thus we can write

$$L_+ U_{\lambda m} = a_{\lambda m}^+ U_{\lambda m+1} \qquad (3.50)$$

and

$$L_- U_{\lambda m} = a_{\lambda m}^- U_{\lambda m-1}. \qquad (3.51)$$

Since the quantity

$$\int U_{\lambda m}^\dagger (L_x^2 + L_y^2)U_{\lambda m}d\Omega \geq 0$$

is a non-negative real number and using Eq. (3.35), Eq. (3.46), Eq. (3.47) and the normalization condition (3.48), we have

$$\int U_{\lambda m}^\dagger (L_x^2 + L_y^2)U_{\lambda m}d\Omega = \int U_{\lambda m}^\dagger (L^2 - L_z^2)U_{\lambda m}d\Omega = \hbar^2(\lambda - m^2) \geq 0.$$

Therefore, $m^2 \leq \lambda$, in other words the number the eigenvalues m in finite. This means that there must be two extrema values m_1 and m_2 such that, according to Eq. (3.50) and Eq. (3.51), we must have

$$L_+ U_{\lambda m_1} = 0 \tag{3.52a}$$

and

$$L_- U_{\lambda m_2} = 0. \tag{3.52b}$$

On the other hand, using Eq. (3.42), the action of $L_+ L_-$ and $L_- L_+$ on the function $U_{\lambda m}$ can be calculated as

$$L_+ L_- U_{\lambda m} = ([L_+, L_-] + L_- L_+) U_{\lambda m} = (2\hbar L_z + L_x^2 + L_y^2 - \hbar L_z) U_{\lambda m}$$
$$= (\hbar L_z + L^2 - L_z^2) U_{\lambda m} = \hbar^2 (\lambda - m^2 + m) U_{\lambda m} \tag{3.53a}$$

and similarly

$$L_- L_+ U_{\lambda m} = \hbar^2 (\lambda - m^2 - m) U_{\lambda m}. \tag{3.53b}$$

Thus, for the eigenvalues m_2 and m_1, if follows that

$$L_+ L_- U_{\lambda m_2} = 0 \tag{3.54a}$$

and

$$L_- L_+ U_{\lambda m_1} = 0. \tag{3.54b}$$

Due to Eq. (3.53), we get the algebraic equations

$$\lambda + m_2 - m_2^2 = 0 \tag{3.54c}$$

and

$$\lambda - m_1 - m_1^2 = 0. \tag{3.54d}$$

The solutions of these equations brings about two possibilities: $m_1 = m_2 - 1$ or $m_1 = -m_2$. The first is not possible since we have chosen $m_1 > m_2$. Therefore, the second solution, $m_1 = -m_2$, is the only possibility. Now, from Eq. (3.50) and Eq. (3.52), we know that the difference $m_1 - m_2$ is a integer number (we can go up by applying step by step the operator L_+ to a given function $U_{\lambda m}$). For this reason $2m_1 \in \mathbf{Z}$. Defining this maximum value m_1 as j, $(m_1 = j)$, the possible values of j are

$$j = 0, \pm 1/2, \pm 1, \pm 3/2, \pm 2, \ldots \tag{3.55a}$$

and from Eq. (3.54d), the eigenvalues λ is given as

$$\lambda = j(j+1). \tag{3.55b}$$

On the other hand, since $\lambda \geq 0$, one must have that $j(j+1) \geq 0$. So, one finally gets its allowed values

$$j = 0, 1/2, \pm 1, \pm 3/2, \pm 2, \ldots \tag{3.55c}$$

The function $U_{\lambda m}$ can be characterized as U_{jm}, where the maximum value of m is $m = j$, and its minimum value is $m = -j$. Let us now calculate the coefficients a_{jm}^+ and a_{jm}^- appearing in Eq. (3.50) and Eq. (3.51). Using Eq. (3.48) and Eq. (3.50), we have

$$1 = \int U_{jm+1}^\dagger U_{jm+1} d\Omega = \frac{1}{|a_{jm}^+|^2} \int (L_+ U_{jm})^\dagger (L_+ U_{jm}) d\Omega$$

$$= \frac{1}{|a_{jm}^+|^2} \int U_{jm}^\dagger (L_- L_+ U_{jm}) d\Omega.$$

Thus, the coefficient a_{jm}^+ can be calculated from the equation

$$|a_{jm}^+|^2 = \int U_{jm}^\dagger (L_- L_+ U_{jm}) d\Omega.$$

Now, using Eq. (3.53b) and Eq. (3.55b) above, it follows

$$|a_{jm}^+|^2 = \hbar^2 (j(j+1) - m(m+1)) = \hbar^2 (j-m)(j+m+1). \tag{3.56a}$$

In a similar way, using the expression

$$1 = \int U_{jm-1}^\dagger U_{jm-1} d\Omega,$$

we get

$$|a_{jm}^-|^2 = \hbar^2 (j(j+1) - m(m-1)) = \hbar^2 (j+m)(j-m+1). \tag{3.56b}$$

We usually consider the coefficients a_{jm}^+ and a_{jm}^- as real numbers. The resulting equations that we obtain from Eq. (3.55), Eq. (3.56), Eq. (3.46) and Eq. (3.47) are the following

$$L_+ U_{jm} = \hbar \sqrt{(j-m)(j+m+1)} U_{jm+1}, \tag{3.57a}$$

$$L_- U_{jm} = \hbar \sqrt{(j+m)(j-m+1)} U_{jm-1}, \tag{3.57b}$$

$$L_x U_{jm} = \frac{\hbar}{2} \left[\sqrt{(j-m)(j+m+1)} U_{jm+1} + \hbar \sqrt{(j+m)(j-m+1)} U_{jm-1} \right], \tag{3.58a}$$

$$L_y U_{jm} = \frac{\hbar}{2i} \left[\sqrt{(j-m)(j+m+1)} U_{jm+1} - \sqrt{(j+m)(j-m+1)} U_{jm-1} \right], \tag{3.58b}$$

$$L^2 U_{jm} = \hbar^2 j(j+1) U_{jm}, \qquad (3.59a)$$

$$L_z U_{jm} = \hbar m U_{jm}, \qquad (3.59b)$$

having the following conditions

$$L_+ U_{j,j} = 0 \qquad (3.60a)$$

and

$$L_- U_{j,-j} = 0. \qquad (3.60b)$$

The allowed values for the quantum number j are given by (3.55c), and the quantum number m must be such that

$$-j \le m \le j. \qquad (3.61)$$

Therefore, for each j value, there are $2j + 1$ possible values for the quantum number m (there are $2j + 1$ possible projections of the angular momentum in the z-axis). Due to Eq. (3.57a) and Eq. (3.58b), we see that it is enough to know a single function U_{jm} to find the set of functions $\{U_{j,-j}, U_{j-j+1}, ..., U_{j,j}\}$ through the application of the operators L_+ and L_- to this given function. Given the eigenvalues j, the easiest functions to obtain are $U_{j,j}$ or $U_{j,-j}$ because they satisfy Eq. (3.60). For example, using Eq. (3.60b) and the operator (3.45), $U_{j,-j}$ is the solution of the following linear partial differential equation

$$-\frac{\partial U_{j,-j}}{\partial \theta} + i \cot \theta \frac{\partial U_{j,-j}}{\partial \phi} = 0. \qquad (3.62)$$

The equation for the characteristics are

$$\frac{d\theta}{-1} = \frac{d\phi}{i \cot \theta} = \frac{dU_{j,-j}}{0}.$$

From the first two terms, we get

$$\int \cot \theta d\theta = i \int d\phi,$$

and making the integration this

$$\log \sin \theta - i\phi = constant.$$

Then, the characteristic curve can be written as

$$c = e^{-i\phi} \sin \theta.$$

The solution of Eq. (3.62) is an arbitrary function, G_j, of the above characteristic curve,

$$U_{j,-j} = G_j(e^{i\phi} \sin \theta). \qquad (3.63)$$

This solution in completaly general for any $2j \in \mathcal{Z}$. However, for $j = l \in \mathcal{Z}$ one can choose a polynomial of order l as a particular solution $(G_l(c) = c^l)$, i.e.

$$U_{l,-l} = a(e^{-i\phi} \sin \theta)^l = ae^{-il\phi} \sin^l \theta,$$

where "a" is a constant of proportionality determined by the normalization condition (3.48),

$$1 = \int U_{l,-l}^{\dagger} U_{l,-l} d\Omega = |a|^2 \int_0^{2\pi} \int_0^{\pi} \sin^{2l+1} \theta d\theta d\phi$$
$$= |a|^2 2\pi \frac{2^{2l+1}(l!)^2}{(2l+1)!}.$$

Hence, the value of the constant of proportionality is given by

$$a = \left[\frac{(2l+1)!}{2^{2l+1} 2\pi (l!)^2} \right],$$

and the function $U_{l,-l}$ is

$$U_{l,-l} = \frac{1}{2^l l!} \sqrt{\frac{(2l+1)!}{4\pi}} e^{-il\phi} \sin^l \theta. \qquad (3.64)$$

Any other function U_{lm} can be calculated now by applying $l + m$ times the operator L_+ to the function $U_{l,-l}$. Using Eq. (3.57a) each time, we obtain the functions

$$U_{lm} = \frac{(-1)^{l+m}}{2^l l!} \sqrt{\frac{(2l+1)!(l-m)!}{4\pi(l+m)!}} e^{im\phi} \sin^m \theta \left[\frac{d^{l+m} \sin^{2l} \theta}{d(\cos \theta)^{l+m}} \right] \qquad (3.65)$$

which are called "Spherical Harmonic Functions", and the usual notation for these functions is $Y_{lm}(\theta, \phi)$ or $Y_l^m(\theta, \phi)$.

3.3 Heat Propagation between Two Superconducting Cables

A superconductor cable (scc) is characterized by a bounded surface containing the origin defined in the space formed by the current (I) carried by the scc (Cooper pairs of electrons), the magnetic field (H) applied to scc, and the temperature (T) of the medium the scc is imbibed. Below this surface the resistivity (ρ) of the scc is zero, and there is not dissipation of energy (heat) generated by flow the current in the cable. Above this surface, the scc behaves like a normal cable whit a very high resistivity value. The surface is called the "critical surface of the scc", and each point on this surface (T_c, I_c, H_c) is called a "critical point of the scc" (see Fig. 3.1).

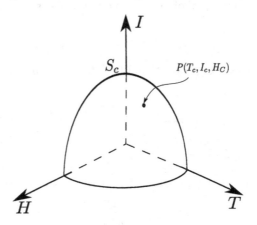

Fig. 3.1   Critical surface and a critical point.

When a region of a scc becomes normal (i.e. this region is above the critical surface) because of some external perturbation, this normal zone propagates along the scc, and the heat generated propagates transversely to other scc. This phenomenon is known as "quenching (or quench)" on the scc. Normally, a scc is made up of many superconducting wires (for example NbTi) contained in a non superconducting matrix (usually Cu or Al). The function of the non superconducting matrix is to make the current flow within it when a quench occurs (because of its much lower resistivity than that of superconducting wires at low temperature) in the scc, avoiding the evaporation of the superconducting wires. Once the quench starts in a scc, the increasing of the temperature where the normal zone first appeared (which maybe the hottest point) as a function of time can be described by the equation

$$(\delta c_p)\frac{dT_1}{dt} = \rho j^2, \qquad (3.66)$$

where (δc_p) is the average heat capacity of the scc, ρ is the average resistivity, and j is the density of the current flowing in the cable. In between two scc there is an insulation. For this reason the transversal propagation of heat is much slower. An approximate description of this transversal heat propagation can be considered as follows: Let two scc's be separated by a distance L and suppose we know the variation of the temperature as a function of time $(T_1(t))$ of the first scc where the quench appeared. At any point "x" between these scc's, the density flow of heat in a distance interval $[x, x + \Delta x]$ is given by qV_h, where q is the density of heat, and $V_h = \Delta x / \Delta t$ is the heat velocity. This heat flow is proportional to the change of the temperature $(\Delta \theta)$ in this interval , and we assume it is inversely proportional to the distance to the scc (x), i.e. $-qV_h = k\Delta\theta / x$, where $k = k(\theta)$ is the function of proportionality, called "thermal conductivity". Therefore, one gets the relation

$$-\Delta q = \frac{k(\theta)}{x} \frac{\Delta\theta}{V_h}. \tag{3.67}$$

On the other hand, the variation of the density of heat in this region is given by

$$\Delta q = (\delta c_p)_I \Delta\theta, \tag{3.68}$$

where $(\delta c_p)_I(\theta)$ is the average heat capacity of the materials in between the two scc's which depends of the temperature. From Eq. (3.68), we get

$$(\delta c_p)_I \frac{\Delta\theta}{\Delta t} + \frac{k}{x} \frac{\Delta\theta}{\Delta x} = 0. \tag{3.69}$$

Taking the limits as Δx and Δt tend to zero, we obtain a quasi-linear partial differential equation for our heat propagation model

$$(\delta c_p)_I(\theta) \frac{\partial\theta}{\partial t} + \frac{k(\theta)}{x} \frac{\partial\theta}{\partial x} = 0. \tag{3.70}$$

The equation for the characteristic of this equation are

$$\frac{dt}{(\delta c_p)_I(\theta)} = \frac{dx}{k(\theta)/x} = \frac{d\theta}{0}.$$

From the last term, we find that one the characteristic curves is

$$C_1 = \theta(x, t),$$

and the substituting this in the other terms, it follows that

$$\frac{dt}{(\delta c_p)_I(C_1)} = \frac{dx}{k(C_1)/x}.$$

We can integrate this last equation obtaining a constant (C_2), characterizing the second characteristic curve, which depends on the fist characteristic (C_1),

$$t - \frac{(\delta c_p)_I(C_1)}{2k(C_1)} x^2 = C_2(C_1)$$

or

$$t - \frac{(\delta c_p)_I(\theta)}{2k(\theta)} x^2 = C_2(\theta). \tag{3.71}$$

C_2 can be determined through the boundary conditions, and this one can be determined in the following way: we know through Eq. (3.66) how the temperature varies in the first scc, that is, at $x = 0$. Therefore, we may impose the boundary condition

$$\theta(0, t) = T_1(t). \tag{3.72}$$

Using this boundary condition in Eq. (3.71), we have

$$t = C_2(T_1(t)),$$

for all time t. But this implies that C_2 is inverse function of T_1, and then Eq. (3.71) looks like

$$t - \frac{(\delta c_p)_I(\theta)}{2k(\theta)} x^2 = T_1^{-1}(\theta).$$

Applying T_1 to this equation, it follows that

$$T_1 \left(t - \frac{(\delta c_p)_I(\theta)}{2k(\theta)} x^2 \right) = \theta. \tag{3.73}$$

The solution of Eq. (3.70) is given in the implicit form Eq. (3.73). If we know the time at which the first scc quenches , it is of practical interest to estimate the time that the heat will take to arrive to the other scc, inducing, consequently, a quench. Suppose that these two scc are separated by a distance L, and the temperature in the second scc as a function of time is $T_2(t)$, then, according to Eq. (3.73), this temperature is given by

$$T_1 \left(t - \frac{(\delta c_p)_I(T_2(t))}{2k(T_2(t))} L^2 \right) = T_2(t). \tag{3.74}$$

If τ_1 is the time taken for the first scc to reach the critical temperature T_{c1} and quench ($T_1(\tau_1) = T_{c1}$), and if τ_2 is the time taken for the second scc to reach the critical temperature T_{c2} and quench ($T_2(\tau_2) = T_{c2}$). Thus, evaluating Eq. (3.74) at the time τ_2, we get

$$T_1 \left(\tau_2 - \frac{(\delta c_p)_I(T_{c2})}{2k(T_{c2})} L^2 \right) = T_{c2},$$

where, in general, $T_{c1} \neq T_{c2}$ since both scc are placed in regions where the magnetic field may by different. However, normally both scc's are close enough to consider that the magnetic fields ere approximately the same, then, we can consider $T_{c1} = T_{c2}$. In this way the right side of the above equation represents the critical temperature of the first scc which happens at the time τ_1. Therefore, it follows that

$$\tau_2 - \frac{(\delta c_p)_I(T_{c2})}{2k(T_{c2})}L^2 = \tau_1$$

or

$$\tau_2 = \tau_1 + \frac{(\delta c_p)_I(T_{c2})}{2k(T_{c2})}L^2. \tag{3.75}$$

The time τ_1 can be estimated, integrating Eq. (3.66), as

$$\tau_1 = \frac{1}{j^2}\int_{T_0}^{T_{c1}} \frac{(\delta c_p)_c(T)}{\rho(T)}dT,$$

where T_0 is the batch temperature.

3.4 Classical Statistical Mechanics in Equilibrium

As we saw in Eq. (1.82), Liouville's theorem gives us the basic equations to deal with many body problems in non equilibrium and in equilibrium situations. To describe ensembles of particles in equilibrium one has to solve the equation

$$\{\rho, H\} = \sum_{i=1}^{3N}\left(\frac{\partial \rho}{\partial q_i}\frac{\partial H}{\partial p_i} - \frac{\partial \rho}{\partial p_i}\frac{\partial H}{\partial q_i}\right) = 0. \tag{3.76}$$

The equations for its characteristic are

$$\frac{dq_1}{\dfrac{\partial H}{\partial p_1}} = ... = \frac{dq_{3N}}{\dfrac{\partial H}{\partial p_{3N}}} = \frac{dp_1}{-\dfrac{\partial H}{\partial q_1}} = ... = \frac{dp_{3N}}{-\dfrac{\partial H}{\partial q_{3N}}} = \frac{d\rho}{0}. \tag{3.77}$$

These equations can be arranged to have

$$\left(\frac{\partial H}{\partial q_1}\right)dq_1 + ... + \left(\frac{\partial H}{\partial q_{3N}}\right)dq_{3N} + \left(\frac{\partial H}{\partial p_1}\right)dp_1 + ... + \left(\frac{\partial H}{\partial p_{3N}}\right)dp_{3N} = 0.$$

But this expression is just

$$dH = 0.$$

Then, we have a characteristic curve given by

$$C = H(\mathbf{q}, \mathbf{p}),$$

and the general solution of Eq. (3.76) is given by any arbitrary function of this characteristic

$$\rho = \rho(H). \tag{3.78}$$

An usual ensemble of particle is the so called "canonical ensemble" defined by

$$\rho = Z^{-1} e^{-\beta H}, \tag{3.79}$$

where Z^{-1} is a normalization constant called "partition function", deduced from

$$\int \rho(\mathbf{q}, \mathbf{p}) \prod_{i=1}^{3N} dp_i dq_i = 1 \tag{3.80}$$

and is given by

$$Z = \int e^{-\beta H} \prod_{i=1}^{3N} dp_i dq_i. \tag{3.81}$$

The factor β is given in terms of the temperature (T) and the Boltzmann's constant (k) as $\beta = 1/kT$. The above expression must have the correction factor $h^{3N} N!$ coming from the Quantum Statical Mechanics. The factor $N!$ takes into account the degenerations due to having undistinguished particles, and the factor h^{3N} is necessary in order for Eq. (3.81) to have the right units and represents the minimum volume which can be occupied in the phase space by the N particles. Thus, Eq. (3.79) and Eq. (3.81) are written as

$$\rho(\mathbf{q}, \mathbf{p}) = \frac{1}{N! h^{3N}} \frac{e^{-\beta H}}{Z} \tag{3.82a}$$

and

$$Z = \frac{1}{N! h^{3N}} \int e^{-\beta H} \prod_{i=1}^{3N} dp_i dq_i. \tag{3.82b}$$

The acknowledge of the partition function (3.82b) allows us know the thermodynamics properties of many particle dynamical system, see reference K. Huang.

Example 3.5. For the example 6 given in first chapter, integrate Eq. (1.89) and corroborate the result (3.78).
Eq. (1.89) is given as

$$\frac{\partial \rho}{\partial q} + \frac{\alpha p}{m} \frac{\partial \rho}{\partial p} = 0. \tag{3.83}$$

The equations for its characteristics are

$$dq = \frac{dp}{(\alpha/m)p} = \frac{d\rho}{0}.$$

From the first two terms, we get the characteristic curve

$$C = pe^{-\alpha q/m},$$

and the Hamiltonian is given in terms of this characteristic curve Eq. (1.86) as

$$H(q,p) = \frac{1}{2m}C^2 = \frac{p^2}{2m}\,e^{-2\alpha q/m}.$$

Then, the solution of Eq. (3.83) can be written as

$$\rho(q,p) = G\left(\frac{p^2}{2m}\,e^{-2\alpha q/m}\right),$$

where G is an arbitrary function.

3.5 Renormalization Group's Equations

If any field theory, a bare vertex is inevitably renormalized by higher order correction diagrams (see Fig. 3.2), and these depend on the momentum of the particle coming into the vertex.

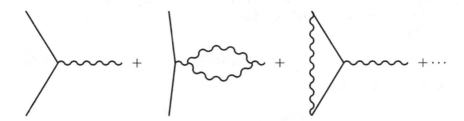

Fig. 3.2   Pair creation and some higher orders diagrams.

This means that we must introduce an effective momentum-dependent vertex,

$$-i\Upsilon^\mu T_{ij}g(\mu)|_{p_i^2} = -\mu^2. \tag{3.84}$$

In calculating higher order diagrams, we generally encounter divergences, for example, in the scalar field theory the loop shown in Fig. 3.3 is logarithmically divergent, as the loop momentum k goes to infinity. This fact can be seen from the integral associated to this loop,

$$\int \frac{dk^4}{(2\pi)^4} \frac{\lambda^2}{(p/2+k)^2(p/2-k)^2}. \tag{3.85}$$

Fig. 3.3   Scalar loop.

One way the renormalize the theory (i.e. to the eliminate all the infinite) is introducing a momentum scale parameter, which depends on choice. Different parameterizations are related by the renormalization group equations

$$\mu\frac{d\Gamma_{n_B n_F}}{d\mu} = -(n_B\Upsilon_B + n_F\Upsilon_F)\Gamma^R_{n_B n_F}(p_i, g, \mu), \qquad (3.86)$$

where $\Gamma^R_{n_B n_F}(p_i, g, \mu)$ is the renormalized Green function which is related to the cut-off Green function $\Gamma^U_{n_B n_F}(p_i, g_0, \Lambda)$ as

$$\Gamma^R_{n_B n_F}(p_i, g, \mu) = [Z_B(\Lambda/\mu)]^{n_B}\, [Z_F(\Lambda/\mu)]^{n_F}\, \Gamma^U_{n_B n_F}(p_i, g_0, \Lambda), \qquad (3.87)$$

where Z_B and Z_F are dimensionless factors, n_B and n_F can be 0,1 or 2, Λ is the cut-off momentum in all the integrations (like Eq. (3.85)) and is introduced by hand to obtain finite quantities as $\Gamma^U_{n_B n_F}(p_i, g_0, \Lambda)$, g_0 is called the bare gauge coupling. The renormalized coupling constant g is given by

$$g(\mu) = Z_B Z_F \Gamma^U_{1,2}. \qquad (3.88)$$

Υ_B and Υ_F are defined as

$$\Upsilon_B(\mu) = -\lim_{\Delta\to\infty}\mu\frac{\partial\log Z_B}{\partial\mu}$$

and

$$\Upsilon_F(\mu) = -\lim_{\Delta\to\infty}\mu\frac{\partial\log Z_F}{\partial\mu}.$$

The Green function expresses the so called "*propagator*" between two vertexes of a Feynman's diagram. The effect of a change in the scale of external momenta in the renormalized Green functions is expressed by

$$\Gamma^R = \Gamma^R_{n_B n_F}(\lambda, p_i, g, \mu) = \mu^D f(\lambda^2 p_i p_j/\mu^2), \qquad (3.89a)$$

or equivalently by the scaling equation

$$\left(\lambda\frac{\partial}{\partial\lambda} + \mu\frac{\partial}{\partial\mu}\right)\Gamma^R = D\Gamma^R, \qquad (3.89b)$$

where D is a constant which tell us the degree of homogeneity. Combining the scaling equation (3.89b) and the renormalization group equation (3.86), we obtain the following PDEFO

$$-\lambda\frac{\partial\Gamma^R}{\partial\lambda} + \beta(g)\frac{\partial\Gamma^R}{\partial g} = -\Delta(g)\Gamma^R, \qquad (3.90)$$

where $\Delta(g)$ and $\beta(g)$ are given by

$$\Delta(g) = D + n_B\Upsilon_B(g) + n_F\Upsilon_F(g) \qquad (3.91a)$$

and

$$\beta(g) = \mu(g\frac{\partial g(\mu)}{\partial\mu}). \qquad (3.91b)$$

In addition, the following condition

$$\Gamma^R(1, g) = G^R(p_i, g), \qquad (3.91c)$$

must be satisfied since for $\lambda = 1$ it is known the expression for the coupling constant and the Green function. The characteristic curves of Eq. (3.90) are found solving the equations

$$\frac{d\lambda}{-\lambda} = \frac{dg}{\beta(g)} = \frac{d\Gamma^R}{-\Delta(g)\Gamma^R}. \qquad (3.92)$$

From the first two terms, we get the characteristic curve given by

$$C = \lambda\exp\left(\int\frac{dg}{\beta(g)}\right), \qquad (3.93)$$

and from the last two terms, one can obtain

$$\Gamma^R = K(C)\exp\left(-\int\frac{\Delta(g)}{\beta(g)}dg\right) \qquad (3.94)$$

or

$$\Gamma^R(\lambda, g) = K(\lambda\psi(g))\phi(g), \qquad (3.95)$$

where $\psi(g)$ and $\phi(g)$ are given by

$$\psi(g) = \exp\left(\int\frac{dg}{\beta(g)}\right) \qquad (3.96a)$$

and

$$\phi(g) = \exp\left(-\int\frac{\Delta(g)}{\beta(g)}dg\right). \qquad (3.96b)$$

$K(C)$ is an arbitrary function of the characteristic curve (3.93) and is determined by the condition (3.91c). Using this condition in Eq. (3.95), we have

$$G^R(p_i, g) = K(\lambda\psi(g))\phi(g)$$

that is

$$K(s) = \frac{G^R\left(p_i, \psi^{-1}(s)\right)}{\phi\left(\psi^{-1}(s)\right)}$$

or

$$K(\lambda\psi(g)) = \frac{G^R\left(p_i, \psi^{-1}(\lambda\psi(g))\right)}{\phi\left(\psi^{-1}(\lambda\psi(g))\right)}.$$

Using this expression in Eq. (3.95), it follows

$$\Gamma^R(\lambda, g) = G^R\left(p_i, \psi^{-1}(\lambda\psi(g))\right) \frac{\phi(g)}{\phi\left(\psi^{-1}(\lambda\psi(g))\right)}. \qquad (3.97)$$

It is of particular interest for physicists to know what happens when $\lambda \to \infty$. According to Eq. (3.93) and Eq. (3.96a), we have

$$\lim_{\lambda\to\infty} \psi(g) = \lim_{\lambda\to\infty} \frac{C}{\lambda} = 0. \qquad (3.98a)$$

Then, this implies from Eq. (3.96a) that

$$\lim_{\lambda\to\infty} \int \frac{dg}{\beta(g)} \to -\infty, \qquad (3.98b)$$

for any g. In particular this must happen for $g = 0$. Therefore, Eq. (3.98b) tell us that the function $\beta(g)$ can be of the form

$$\beta(g) = a_n g^n, \qquad (3.99)$$

where a_n is a constant and n can be $1, 2, 3, \dots$. This happens for example to the diagrams of Fig. 3.4, where $\beta(g)$ is expressed as $\beta(g)E - g^3 + O(> 3)$. From left to right in this figure, the first loop corresponds to gluons, the second one to ghosts, and the last one to quarks. This expresses what is called "asymptotic freedom" for Quantum Cromo Dynamics (QCD). A complete discussion about this can be found in reference of J.Ellis and C.T. Sanchrajda.

3.6 Particle Multiplicity Distribution in High Energy Physics

Collisions of two beam of particles (proton (p)-antiproton ($\bar{p}$); electron (e^-)-positron (e^+); etc.) at high energies, from 10 GeV up to around 900 GeV (1GeV= 10^9 eV) produces more new charged ($\pm$) or neutral (o) particles such as pions ($\pi^{\pm,o}$), Kaons ($K^{\pm}$), Lambdas (Λ^o), Sigmas ($\sum^{\pm,o}$), etc., which are called "secondary particles". The amount of secondary particles

Fig. 3.4

produced is called the "multiplicity of the event". If our detector is capable of knowing all the particles appearing after the collision, one talks about "exclusive events detector". If our detector can only know with detail few of these outgoing particles, one talks about of an "inclusive events detector." One is interested in to find, for a given energy collision, the number of charged particles (or neutral particles) which can appear after the collision, i.e. the multiplicity. Looking at the problem in a different way, we want to find the probability P_n of finding a number "n" of charged particles for a given energy. A possible probability distribution is given in the reference of P. Srivastava,

$$P_n = e^{-<n>} \frac{B^n}{n!} A_n(k \exp(-B)),\qquad(3.100)$$

where $< n >$ denote the average number of charged particles, k is referred to the k-th moment $< x^k >$, B is defined as $B =< n > /k$, and $A_n(x)$ are the polynomials defined by the recurrent relation

$$A_{n+1} = x \left[A_n(x) + \frac{dA_n(x)}{dx} \right]\qquad(3.101)$$

satisfying the initial condition

$$A_0(x) = 1.\qquad(3.102)$$

One approximation that we can make to Eq. (3.100) is by assuming that the variable "n" changes continuously. Thus, the term of Eq. (3.101) can be taken at first order approximation in the Taylor's series as

$$A_{n+1}(x) = A_n(x) + \frac{\partial A_n(x)}{\partial n}.\qquad(3.103)$$

In his way, making use of Eq. (3.103) and Eq. (3.101), and after rearranging terms, a PDEFO is obtained

$$\frac{\partial A}{\partial n} - x\frac{\partial A}{\partial x} = (x - 1)A,\qquad(3.104)$$

where we have defined $A = A(n, x) = A_n(x)$. Eq. (3.102) is

$$A(0, x) = 1. \tag{3.105}$$

The equation for the characteristic of Eq. (3.104) are given by

$$\frac{dn}{1} = \frac{dx}{-x} = \frac{dA}{(x-1)A}. \tag{3.106}$$

From the first two term, we obtain the equation

$$dn = \frac{dx}{-x} \tag{3.107}$$

which can be integrated, obtaining the characteristic curve

$$C_1 = n + \log x. \tag{3.108}$$

The characteristic curve can be chosen as $C = \exp(C_1)$, that is

$$C = xe^n. \tag{3.109}$$

From the last two terms of Eq. (3.106), we have the equation

$$\frac{dx}{-x} = \frac{A}{(x-1)A} \tag{3.110}$$

which can be written as

$$\int \frac{(x-1)dx}{x} = \int \frac{dA}{A} \tag{3.111}$$

and can be integrated, obtaining the solution

$$A = K(C)xe^{-x}, \tag{3.112}$$

or using Eq. (3.109), it follows that

$$A(n, x) = K(xe^n)xe^{-x}. \tag{3.113}$$

with the condition (3.105) in Eq. (3.113), we have

$$1 = K(x)xe - x,$$

or

$$K(s) = s^{-1}e^s.$$

Then, the function K is determined as

$$K(xe^n) = x^{-1}e^n e^{xe^n}, \tag{3.114}$$

and the solution of Eq. (3.104) is finally given as

$$A(n, x) = e^{-(x+n)} \exp(x \exp(n)). \tag{3.115}$$

3.7 Hamiltonian Perturbation Approach in Accelerator Physics

Accelerator Physics studies the behavior of a charged particle inside an accelerator machine. Although there are many accelerator structures, the one of particular interest for applications is the synchrotron machine where the charged particles travel inside a ring, and the trajectories are determined mainly by the magnetic field, **B**, provided by the ideal multipole magnets placed all over the ring (dipole magnets are use for bending, quadrupole magnets are use for focusing, sextupole magnets are use for chromaticity corrections, etc.). These ideal (not constructions errors are assumed) magnets define the so called "lattice" of the synchrotron machine. The transverse behavior of the charged particle in this structure can be described using the Hamiltonian approach where the Hamiltonian is given in reference E.D. Courant and H.S. Snyder, and the reference M. Sand,

$$
H = \frac{1}{2}\left(P_x - \frac{e}{cp}A_x\right)^2 + \frac{1}{2}\left(P_y - \frac{e}{cp}A_y\right)^2
$$
$$
- \frac{e}{cp}A_s + \frac{1}{2}K_1(s)y^2 - \frac{1}{2}\left(K_1(s) - \frac{1}{\rho^2}\right)x^2, \quad (3.116)
$$

where p is the longitudinal linear momentum of the charged particle, $P_x = p_x/p$ and $P_y = p_y/p$ are its normalized transversal momenta, e is the charged of the particle, c is the speed of light, $\rho(s)$ it the curvature of the accelerator ring, $K_1(s)$ describes the linear lattice (ideal dipoles and quadrupoles) of the machine, and the vector potential, $\mathbf{A} = (A_x, A_y, A_s)$, represents nonlinear magnets or additional electromagnetic field due to other possible sources (nonlinear components of the real linear magnets, or longitudinal multipole periodic pattern along the real magnets, for example). Separating the nonlinear term of the vector potential, this Hamiltonian can be written as

$$
H(\mathbf{x}, \mathbf{P}, s) = H_0(\mathbf{x}, \mathbf{P}, s) + V(\mathbf{x}, \mathbf{P}, s), \quad (3.117)
$$

where $\mathbf{x}$ and $\mathbf{P}$ are the vectors $\mathbf{x} = (x, y)$ and $\mathbf{P} = (P_x, P_y)$, and H_0 and V are given by

$$
H_0(\mathbf{x}, \mathbf{P}, s) = \frac{1}{2}(P_x^2 + K_x(s)x^2) + \frac{1}{2}(P_y^2 + K_y(s)y^2) \quad (3.118a)
$$

and

$$
V(\mathbf{x}, \mathbf{P}, s) = -\frac{e}{cp}(A_s + P_x A_x + P_y A_y). \quad (3.118b)
$$

Note that the description is done taking the length of the accelerator ring "s" as the independent parameter. It is convenient to make an action-angle variables canonical transformation through the generating function

$$F(\mathbf{x}, \phi, s) = -\sum_{i=1}^{2} \frac{x_i^2}{2\beta_i(s)} (\tan(\phi_i) - \beta_i'(s)/2), \quad \text{for} \quad i = 1, 2 \qquad (3.119)$$

where x is for $i = 1$ and y is for $i = 2$, $\beta_i'(s)$ is the derivative with respect to s of the beta function $\beta_i(s)$ which satisfies the differential equation

$$2\beta_i\beta_i'' - (\beta_i')^2 + 4K_i(s)\beta_i^2 = 4, \qquad (3.120a)$$

and the $\phi_i(s)$ is called the betatron phase which is related with the beta function through

$$\phi_i(s) = \phi_i(0) + \int_0^s \frac{d\sigma}{\beta_i(\sigma)}. \qquad (3.120b)$$

The action, I_i, the coordinates, and the canonical linear momenta are given by

$$I_i = -\frac{\partial F}{\partial \phi_i} = \frac{1}{2\beta_i(s)} [x_i^2 + (\beta_i x_i' - \beta_i' x_i/2)], \qquad (3.121a)$$

$$x_i = \sqrt{2I_i\beta_i} \cos(\phi_i), \qquad (3.121b)$$

and

$$P_i = -\sqrt{\frac{2I_i}{\beta_i}} \left(\sin(\phi_i) - \frac{1}{2}\beta_i' \cos(\phi_i) \right), \quad \text{for} \quad i = 1, 2. \qquad (3.121c)$$

Therefore, Hamiltonian (3.117) written in terms of the action-angle variables $(\phi, \mathbf{I})$ is given by

$$H(\phi, \mathbf{I}, s) = \sum_{i=1}^{2} \frac{I_i}{\beta_i(s)} + \mathcal{V}(\phi, \mathbf{I}, s), \qquad (3.122)$$

where the function $\mathcal{V}$ is defined as

$$\mathcal{V}(\phi, \mathbf{I}, s) = V(\mathbf{x}(\phi, \mathbf{I}), \mathbf{P}(\phi, \mathbf{I}), s). \qquad (3.123)$$

One the most important parameters that the accelerator physicists are interested in is the so called "tune" of the machine,

$$\nu_i = \frac{1}{2\pi} \int_0^{\mathcal{C}} \frac{d\sigma}{\beta_i(\sigma)}, \quad \text{for} \quad i = 1, 2, \qquad (3.124)$$

which has the meaning of number of betatron oscillations performed by the particle around the ring of length $\mathcal{C}$. If the function $\mathcal{V}$ in Eq. (3.122) is different from zero, this tune is modified (the tune is shifted if the modification is the same for all the particles, or there is "tune spread" if the modification is different for different particles). To calculate this change at first order in perturbation theory, the averages over the phases ($\phi_i \in [0, 2\pi]$, $i = 1, 2$) and the integration over the length of the ring on the Hamiltonian (3.122) is carried out

$$\widetilde{H}(\mathbf{I}) = < H(\phi, \mathbf{I}, s) > = \frac{1}{(2\pi)^3} \int_0^C ds \int_0^{2\pi} \int_0^{2\pi} d\phi_i d\phi_2 H(\phi, \mathbf{I}, s). \quad (3.125)$$

And the modified tune (ν_i') is just the derivative of this averaged Hamiltonian with respect the action

$$\nu_i' = \frac{\partial \widetilde{H}(\mathbf{I})}{\partial I_i}. \quad (3.126)$$

To go to a second-order perturbation, a new canonical transformation close to the identity can be made. This is achieved with the help of the generating function

$$\widetilde{F}(\phi, \mathbf{J}, s) = \sum_{i=1}^2 \phi_i J_i + G(\phi, \mathbf{J}, s). \quad (3.127)$$

The action between the new variables ($\boldsymbol{\Phi}, \mathbf{J}$) and the old one ($\phi, \mathbf{I}$) is given by the expressions

$$I_i = \frac{\partial \widetilde{F}}{\partial \phi_i} = J_i + \frac{\partial G}{\partial \phi_i} \quad (3.128a)$$

and

$$\Phi_i = \frac{\partial \widetilde{F}}{\partial J_i} = \phi_i + \frac{\partial G}{\partial J_i}. \quad (3.128b)$$

Hamiltonian (3.122) written in terms of these new variables is

$$\mathcal{H}(\boldsymbol{\Phi}, \mathbf{J}, s) = \sum_{i=1}^2 \frac{J_i}{\beta_i(s)} + \sum_{i=1}^2 \frac{1}{\beta_i(s)} \frac{\partial G}{\partial \phi_i} + \frac{\partial G}{\partial s} + \widetilde{\mathcal{V}}(\boldsymbol{\Phi}, \mathbf{J}, s), \quad (3.129)$$

where the function $\widetilde{\mathcal{V}}$ is given by

$$\widetilde{\mathcal{V}}(\boldsymbol{\Phi}, \mathbf{J}, s) = \mathcal{V}\left(\boldsymbol{\Phi} - \frac{\partial G}{\partial \mathbf{J}}, \mathbf{J} - \frac{\partial G}{\partial \phi}, s\right). \quad (3.130)$$

Doing a Taylor expansions of this function

$$\widetilde{\mathcal{V}}(\boldsymbol{\Phi}, \mathbf{J}, s) = \mathcal{V}(\boldsymbol{\Phi}, \mathbf{J}, s) + \sum_{i=1}^2 \left(\frac{\partial \mathcal{V}}{\partial J_i} \frac{\partial G}{\partial \Phi_i} - \frac{\partial \mathcal{V}}{\partial \Phi_i} \frac{\partial G}{\partial J_i}\right) + \cdots, \quad (3.131)$$

and neglecting terms of order higher than one, the Hamiltonian would be
written as

$$\mathcal{H}(\mathbf{\Phi}, \mathbf{J}, s) = \sum_{i=1}^{2} \frac{J_i}{\beta_i(s)} + \sum_{i=1}^{2} \left(\frac{\partial \mathcal{V}}{\partial J_i} \frac{\partial G}{\partial \Phi_i} - \frac{\partial \mathcal{V}}{\partial \Phi_i} \frac{\partial G}{\partial J_i} \right), \quad (3.132)$$

if the function G satisfies the following linear partial differential equation

$$\sum_{i=1}^{2} \frac{1}{\beta_i(s)} \frac{\partial G}{\partial \phi_i} + \frac{\partial G}{\partial s} + \mathcal{V}(\mathbf{\Phi}, \mathbf{J}, s) = 0, \quad (3.133)$$

where it has been assumed $\phi \approx \mathbf{\Phi}$. The equation for the characteristics are
given by

$$\beta_1(s)d\phi_1 = \beta_2(s)d\phi_2 = \frac{dG}{-\mathcal{V}(\phi, \mathbf{J}, s)} = ds. \quad (3.134)$$

From the first two terms and the last one of Eq. (3.134), two characteristic
curves follow

$$C_1 = \phi_1 - \psi_1(s) \quad \text{and} \quad C_2 = \phi_2 - \psi_2(s), \quad (3.135)$$

where the function $\psi_i(s)$ is defined as

$$\psi_i(s) = \int_0^s \frac{d\sigma}{\beta_i(\sigma)} \quad \text{for} \quad i = 1, 2. \quad (3.136)$$

Using these equations in third term of Eq. (3.134), it follows that

$$G(\phi, \mathbf{J}, s) = - \int_0^s \mathcal{V}(C_1 + \psi_1(\xi), C_2 + \psi_2(\xi), \mathbf{J}, \xi)d\xi + A(C_1, C_2). \quad (3.137)$$

Choosing the conditions $G(\phi, \mathbf{J}, 0) = 0$, the solution of Eq. (3.133) is given
by

$$G(\phi, \mathbf{J}, s) = - \int_0^s \mathcal{V}(\phi_1 - \psi_1(s) + \psi_1(\xi), \phi_2 - \psi_2(s) + \psi_2(\xi), \mathbf{J}, \xi)d\xi. \quad (3.138)$$

This solution is, then, substituted in Eq. (3.132) to be able to make the
average and calculate the modified tune of the machine through Eq. (3.126)
(for a explicit application see G. López and S. Chen). Assume the perturbation function is periodic on the angles and distance variables. Then, this
function can be written in Fourier expansion as

$$\mathcal{V}(\phi, \mathbf{J}, s) = \sum_{m_1, m_2, l} \mathcal{V}_{m_1, m_2, l}(\mathbf{J}) e^{i(m_1\phi_1 + m_2\phi_2 - l\Omega s)}.$$

Therefore, Eq. (3.138) can be expressed as

$$\mathcal{V}(\phi_1 - \psi_1(s) + \psi_1(\xi), \phi_2 - \psi_2(s) + \psi_2(\xi), \mathbf{J}, \xi) = \sum_{m_1, m_2, l} \mathcal{V}_{m_1, m_2, l}(\mathbf{J})$$

$$\times e^{i[m_1(\phi_1 - \psi_1(s) + \psi_1(\xi)) + m_2(\phi_2 - \psi_2(s) + \psi_2(\xi)) - l\Omega\xi]}, \quad (3.139)$$

and substituting this expression in Eq. (3.138), we obtain

$$G(\phi, \mathbf{J}, s) = \sum_{m_1, m_2, l} \mathcal{V}_{m_1, m_2, l}(\mathbf{J}) e^{i[m_1(\phi_1 - \psi_1(s)) + m_2(\phi_2 - \psi_2(s))]}$$

$$\times \int_0^s e^{i[m_1\psi_1(\xi) + m_2\psi_2(\xi) - l\Omega\xi]} d\xi. \quad (3.140)$$

Suppose now that $\psi(s) \approx \omega_i s$, the integral of the right hand of (3.140) can be expressed as

$$\int_0^s e^{i[m_1\psi_1(\xi) + m_2\psi_2(\xi) - l\Omega\xi]} d\xi = \int_0^s e^{i[m_1\omega_1 + m_2\omega_2 - l\Omega]\xi} d\xi$$

$$= \begin{cases} i\dfrac{1 - e^{i[m_1\omega_1 + m_2\omega_2 - l\Omega]s}}{m_1\omega_1 + m_2\omega_2 - l\Omega} & \text{if } m_1 w_1 + m_2 w_2 \neq l\Omega, \\ \\ s & \text{if } m_1 w_1 + m_2 w_2 = l\Omega. \end{cases} \quad (3.141)$$

The first case ($m_1 w_1 + m_2 w_2 \neq l\Omega$) is called **non resonant case**, and the other one ($m_1 w_1 + m_2 w_2 = l\Omega$) is called **resonant case**. In this latter situation one says that one has **non linear resonances** characterized by the pair (m_1, m_2). Because in the non linear resonant situation the function G grows linearly with the parameter s, there will be a value s^* for which this approximation will not be valid any more since it will have a big value. These non linear resonance are import, for example, for determine the life time of a beam in a circular accelerator, the stability of a planetary system, the chaotic behavior of a diatomic molecule. Given a non linear resonance (m_1, m_2), this defines a rational relation among the involved frequencies, $\omega_1 = -\frac{m_2}{m_1}\omega_2 + \frac{l}{m_1}\Omega$, and since the rational numbers are dense in the real number, any operational point in the space ($\omega_1^{op}, \omega_2^{op}$) of the system will be close enough of a non linear resonance which could make it unstable. Of course, nobody is interested in having a non linear system which last forever. So, one just try to operate the system away of the "danger non linearities" for the machine to have the life time required. For natural periodic systems, one first chooses their operation point, then one tries to determine all non linearities, and finally one tries to make some estimation of their stability lifetime.

The procedure outlined in this section is called Kolmogorov perturbation approach (see reference V.I. Kolmogorov), and note from Eq. (3.137)

that the function G is proportional to the perturbation $\mathcal{V}(\mathbf{\Phi},\mathbf{J},s)$. Therefore, Eq. (3.131) represents the Hamiltonian approximated up to second
order in $\mathcal{V}$. If we call the second term on the right side of Eq. (3.131) as
$\mathcal{V}^{(2)}(\mathbf{\Phi},\mathbf{J},s)$, the resulting expression will have exactly the same form as
Eq. (3.121). Then, one can do another infinitesimal canonical transfor-
mation with a similar generatrix function (3.126) in terms of the variables
$\{\mathbf{\Phi},\mathbf{J},s\}$. So, following the same procedure we have done before with this
Hamiltonian, we gets a similar expressions (3.131) and (3.137) but with
function $\mathcal{V}^{(2)}$ instead of $\mathcal{V}$, that is, Eq. (3.137) would be of second order
in $\mathcal{V}$, and Eq. (3.131) would be of fourth order in $\mathcal{V}$. Thus, in general,
for a nth-procedure, one gets and approximation of 2^n-order in $\mathcal{V}$ with a
resulting Hamiltonian

$$H^{(n)}(\mathbf{\Phi},\mathbf{J},s) = \sum_{i=1}^{2} \frac{J_i}{\beta_i(s)} + \mathcal{V}^{(2^n)}(\mathbf{\Phi},\mathbf{J},s) \, ,$$

where $\mathcal{V}^{(2^n)}$ is given by

$$\mathcal{V}^{(2^n)}(\mathbf{\Phi},\mathbf{J},s) = \sum_{i=1}^{2} \left(\frac{\partial \mathcal{V}^{(2^{n-1})}}{\partial J_i} \frac{\partial G}{\partial \Phi_i} - \frac{\partial \mathcal{V}^{(2^{n-1})}}{\partial \Phi_i} \frac{\partial G}{\partial J_i} \right)$$

and the function G has the following expression

$$G(\mathbf{\Phi},\mathbf{J},s) = - \int_0^s \mathcal{V}^{(2^{n-1})}\left(\Phi_1 - \psi_1(s) + \psi_1(\xi), \Phi_2 - \psi_2(s) + \psi_2(\xi), \mathbf{J}, s\right) d\xi \, .$$

Of course, one must have that the following relations must be satisfied

$$\left| \mathcal{V}^{(2^n)} \right| < \left| \mathcal{V}^{(2^{n-1})} \right| \, , \quad \text{and} \quad \left| \sum_{i=1}^{2} \frac{J_i}{\beta_i(s)} \right| \ll \left| \mathcal{V} \right|$$

for $\Phi_i \in [o, 2\pi]$, $s, J_i \in [0, \infty)$, $n \in \mathcal{Z}^+$. On each procedure there will be a
modification on the tune through the expression

$$\nu_i^{(n)} = \frac{\partial \langle H^{(n)} \rangle}{\partial J_i} \quad i = 1, 2 \, ,$$

with the average defined as Eq. (3.124). If the procedures converge, it does
in an exponential way, 2^n, see also reference V.I. Arnold.

3.8 Perturbation Approach for the One-Dimensional Constant of Motion

In what follows, one is restricted oneself to one-dimensional system for sim-
plicity. As it was shown in first section of this chapter, the constant of

motion of an autonomous system always exists and is given by the solution Eq. (3.3). However, in most cases this equation is quite difficult to solve (specially for nonlinear forces). Due to the importance of knowing this constant of motion, one would like to know how this one is, at least, with some degree of approximation. To do this, the following perturbation method may be used. Suppose that the force can be separated in two parts,

$$f(q, v) = f_0(q, v) + f_I(q, v). \qquad (3.142)$$

The first part, f_0, is the part of the force for which Eq. (3.3) is integrable and has the known solution $K_0(q, v)$,

$$v\frac{\partial K_0}{\partial q} + f_0(q, v)\frac{\partial K_0}{\partial v} = 0,$$

and the second part, f_I, is considered as a perturbation to the system. Suppose also that the constant of motion can be expressed of the form

$$K(q, v) = K_0(q, v) + \sum_{n=1}^{\infty} K^{(n)}(q, v), \qquad (3.143)$$

where $K^{(n)}$ is of the order $(f_I)^n$ in force strength, that is $\mathcal{O}\left(K^{(n)}\right) \sim (f_I)^n$. Defining $K^{(0)} = K_0$ and substituting this expressions in

$$v\frac{\partial K}{\partial q} + [f_0(q, v) + f_I(q, v)]\frac{\partial K}{\partial v} = 0, \qquad (3.144)$$

it follows

$$v\frac{\partial K_0}{\partial q} + f_0(q, v)\frac{\partial K_0}{\partial v} + \sum_{n=1}^{\infty}\left\{f_I(q, v)\frac{\partial K^{(n)}}{\partial v}\right\} +$$

$$\sum_{n=1}^{\infty}\left\{v\frac{\partial K^{(n)}}{\partial q} + f_0(q, v)\frac{\partial K^{(n)}}{\partial v} + f_I(q, v)\frac{\partial K^{(n-1)}}{\partial v}\right\} = 0. \quad (3.145)$$

Neglecting the terms of the order $(f_I)^{n+1}$, the following equation can be derived

$$v\frac{\partial K^{(n)}}{\partial q} + f_0(q, v)\frac{\partial K^{(n)}}{\partial v} = -f_I(q, v)\frac{\partial K^{(n-1)}}{\partial v}, \quad \text{for} \quad n = 1, 2... \quad (3.146)$$

This is a linear partial differential equation of first order, and its equations for the characteristics are

$$\frac{dq}{v} = \frac{dv}{f_0(q, v)} = \frac{dK^{(n)}}{-f_I(q, v)K_v^{(n-1)}}, \qquad (3.147)$$

where $K^{(n-1)}$ means the partial differentiation of $K^{(n-1)}$ with respect to v. From the first two terms of this equation one can obtain the same function K_0 as a characteristic curve of Eq. (3.146) which can be used to get $v = v(K_0, q)$ which, in turn, is used in Eq. (3.147) to get

$$K^{(n)}(q, v) = -\int^q \frac{f_I(\xi, v(K_0, \xi)) K_v^{(n-1)}}{v(K_0, \xi)} d\xi, \qquad (3.148)$$

or applying iteratively this equation, the constant of motion for the autonomous system is given by

$$K(q, v) = K_0(q, v) + \sum_{n=1}^{\infty} \left[(-1)^n \prod_{i=1}^{n} \int^{\xi_{i-1}} d\xi_i \frac{f_I(\xi_i)}{v(K_0, \xi_i)} \frac{\partial}{\partial v} \right] K_0, \quad (3.149)$$

where $\xi_0 = q$ (for application to explicit examples see reference N.M. Atakishiyev et al).

3.9   Constant of Motion for a Relativistic Particle under Periodic Perturbation

The one-dimensional motion of the relativistic particle under the periodic perturbation force can be described by the following Newton equation of motion

$$m\frac{d^2 x}{dt^2} = e \left(1 - \frac{v^2}{c^2} \right)^{3/2} E \sin(kx - \omega t), \qquad (3.150)$$

where e, m and v are the charged, the mass at rest and the velocity of the particle. E, k and ω are the amplitude, wave number and frequency of oscillation of the field, c is the speed of light. Defining the new variables ξ

$$\xi = kx - \omega t \qquad (3.151)$$

Eq. (3.150) can be written as the following autonomous dynamical system

$$\frac{d\xi}{d\tau} = kv - \omega, \qquad (3.152a)$$

$$\frac{dv}{d\tau} = \left(1 - \frac{v^2}{c^2} \right)^{3/2} A \sin\xi, \qquad (3.152b)$$

where the coefficient A is defined as $A = eE/m$. A constant of motion of this system is given by a solution of the following equation

$$(kv - \omega)\frac{\partial K}{\partial \xi} + \left(1 - \frac{v^2}{c^2} \right)^{3/2} A \sin\xi \frac{\partial K}{\partial v} = 0. \qquad (3.153)$$

The equation for the characteristic curves are

$$\frac{d\xi}{kv - \omega} = \frac{dv}{\left(1 - \frac{v^2}{c^2} \right)^{3/2} A \sin\xi} = \frac{dK}{0}. \qquad (3.154)$$

Solving these equations, the constant of motion which the right limit for the non-relativistic case $(c \to \infty)$ is given by

$$K(\xi, v) = \frac{mc^2}{\sqrt{1 - v^2/c^2}} - mc^2 - \frac{m\omega v}{k\sqrt{1 - v^2/c^2}} + \frac{mA}{k}\cos(kx - \omega t). \quad (3.155)$$

3.10 Uniqueness of Constant of Motion

In section 3.1, a criteria for getting a constant of motion for a dissipative system was given. This criteria is based on taking the limit for the dissipative parameter to zero, and looking for the functionality of the arbitrary function such that one gets the know function at this limit (the usual energy of a conservative system). However, this criteria does not guarantee that this constant of motion is unique, in fact, consider the following dynamical quadratic dissipative system under a constant force mf, being m the mass of the particle,

$$\frac{dx}{dt} = v , \qquad \frac{dv}{dt} = f - \frac{\alpha}{m}v^2 , \qquad (3.156)$$

where α is the dissipation parameter, and one has $v \geq 0$. The constant of motion of this autonomous system $K(x, v)$ satisfies the linear PDEFO

$$v\frac{\partial K}{\partial x} + \left(f - \frac{\alpha}{m}v\right)\frac{\partial K}{\partial v} = 0 \qquad (3.157)$$

which has the general solution

$$K(x, v) = G\big(C(x, v)\big) , \quad \text{with} \quad C(x, v) = \frac{mf}{2\alpha}\ln\left(1 - \frac{\alpha v^2}{mf}\right) - fx , \qquad (3.158)$$

where G is an arbitrary function of the characteristic curve "C". By choosing

$$G(C) = mC , \qquad \text{or} \quad G(C) = -\frac{m^2 f}{2\alpha}e^{-2\alpha C/m} + \frac{m^2 f}{2\alpha} \qquad (3.159)$$

one gets the constants of motion

$$K_\alpha^{(1)}(x, v) = -\frac{m^2 f}{2\alpha}\ln\left(1 - \frac{\alpha v^2}{mf}\right) - mfx , \qquad (3.160)$$

or

$$K_\alpha^{(2)}(x, v) = -\frac{m^2}{2\alpha}\left(f - \frac{mv^2}{m}\right)e^{2\alpha x/m} + \frac{m^2 f}{2\alpha} . \qquad (3.161)$$

Using Taylor expansion on both constants of motion, one has that

$$\lim_{\alpha \to 0} K_\alpha^{(i)}(x, v) = \frac{1}{2}mv^2 - mfx , \quad \text{for} \quad i = 1, 2 . \qquad (3.162)$$

Both different constants of motion has the same expected "physical" limit, and this represents a strong ambiguity for Lagrangian, Hamiltonian and applications on quantum and statistical mechanics.

To overcome this ambiguity when dealing with velocity dependent forces, let us recall from chapter 2 that the theorem of existent and uniqueness of a linear and quasilinear PDEFO is stablished whenever the curve of initial data is not one of the characteristic curves. Therefore, assuming that the curve of initial data

$$\Gamma_0 : \left\{ x = 0 \ , \quad K = \frac{1}{2}mv^2 \right\} \tag{3.163}$$

is not a characteristic curve of the linear PDEFO

$$v\frac{\partial K}{\partial x} + \frac{F(x,v)}{m}\frac{\partial K}{\partial v} = 0 \ , \tag{3.164}$$

associated to the constant of motion of the dynamical system

$$\frac{dx}{dt} = v \ , \qquad \frac{dv}{dt} = \frac{F(x,v)}{m} \ , \tag{3.165}$$

the solution will be unique determined (without the ambiguity mentioned before).

Let us apply this idea to the above dynamical system. The general solution is given by Eq. (3.157), and applying the initial condition of Eq. (3.162), $K(0,v) = mv^2/2$, it follows that

$$\frac{1}{2}mv^2 = G\left(-\frac{mf}{2\alpha} \ln\left(1 - \frac{\alpha v^2}{mf}\right) \right) . \tag{3.166}$$

Defining σ as

$$\sigma = -\frac{mf}{2\alpha} \ln\left(1 - \frac{\alpha v^2}{mf}\right) , \tag{3.167}$$

one has that

$$v^2 = \frac{mf}{\alpha}\left(1 - e^{-2\alpha\sigma/mf}\right) \tag{3.168}$$

and

$$G(\sigma) = \frac{m^2 f}{2\alpha}\left(1 - e^{-2\alpha\sigma/mf}\right) . \tag{3.169}$$

Therefore and after some rearrangements, the constant of motion which pass on the above initial curve is given by

$$K_\alpha(x, v) = \frac{m^2 f}{2\alpha} \left[1 - \left(1 - \frac{\alpha v^2}{mf} \right) e^{2\alpha x/m} \right] . \tag{3.170}$$

This constant of motion has the expected limit

$$\lim_{\alpha \to 0} K_\alpha(x, v) = \frac{1}{2} mv^2 - mfx . \tag{3.171}$$

This curve of initial conditions looks to work fine for the non relativistic case ($v \ll c$, with "c" being the speed of light). However, for the relativistic case ($v \lesssim c$) it is better to select the following curve of initial data

$$\Gamma_0 : \left\{ x = 0 ; \quad K = (\gamma - 1)mc^2 | \text{where} \quad \gamma = \left(1 - \frac{v^2}{c^2} \right)^{-1/2} \right\} . \tag{3.172}$$

The reason for doing this is that for a free relativistic particle one knows that the energy is given by $(\gamma - 1)mc^2$, having the right non relativistic limit when $v/c \ll 1$ (that is, $mv^2/2$).

Consider a relativistic particle moving under a conservative force $F(x)$. Its associated dynamical system can be written as

$$\frac{dx}{dt} = v , \qquad \frac{dv}{dt} = \frac{F(x)}{m} \left(1 - \frac{v^2}{c^2} \right)^{3/2} . \tag{3.173}$$

The constant of motion $(K(x, v))$ for this autonomous system is the solution of the following linear PDEFO

$$v \frac{\partial K}{\partial x} + \frac{F(x)}{m} \left(1 - \frac{v^2}{c^2} \right)^{3/2} \frac{\partial K}{\partial v} = 0. \tag{3.174}$$

Its characteristic equations are given by

$$\frac{dx}{v} = \frac{dv}{\frac{F(x)}{m} \left(1 - \frac{v^2}{c^2} \right)^{3/2}} = \frac{dK}{0} \tag{3.175}$$

which has the general solution

$$K(x, v) = G\big(C(x, v)\big) , \tag{3.176}$$

where "C" is the characteristic curve deduced form the first two term of this expression. Separating variables, one arrives to the following integration

$$\int F(x)dx = m \int \frac{vdv}{\left(1 - \frac{v^2}{c^2} \right)^{3/2}} , \tag{3.177}$$

and its integration brings about the characteristic curve

$$C(x, v) = \gamma mc^2 + V(x) \,, \tag{3.178}$$

where $V(x)$ represents the potential energy $(V(x) = -\int F(x)dx)$, and γ is the known function $\gamma(v) = (1 - v^2/c^2)^{-1/2}$. Let us call $V_0 = V(0)$ and apply the initial data to our solution, obtaining the expression

$$(\gamma - 1)mc^2 = G(\gamma mc^2 + V_0) \,. \tag{3.179}$$

Defining now $\sigma = \gamma mc^2 + V_0$, one gets

$$G(\sigma) = \sigma - V_0 - mc^2 \,. \tag{3.180}$$

Having this functionality already, after making the evaluation of G in $\gamma mc^2 + V(x)$, one obtains the solution

$$K(x, v) = (\gamma - 1)mc^2 + V(x) - V_0 \,. \tag{3.181}$$

This constant of motion is the usual relativistic expression for a conservative system with one degree of freedom, has the right limit for the non relativistic case, and is uniquely defined.

Of course, it is not always possible to get $v = v(\sigma)$ during the process of the determination of the function G, for example, consider a relativistic quadratic dissipative system $(v \geq 0)$ under a constant force F, described by the dynamical system

$$\frac{dx}{dt} = v \,, \qquad \frac{dv}{dt} = \left(\frac{F}{m} - \frac{\alpha v^2}{m}\right)\left(1 - \frac{v^2}{c^2}\right)^{3/2} \,. \tag{3.182}$$

The constant of motion for this autonomous system is the solution of the following liner PDEFO

$$v\frac{\partial K}{\partial x} + \left(\frac{F}{m} - \frac{\alpha v^2}{m}\right)\left(1 - \frac{v^2}{c^2}\right)^{3/2}\frac{\partial K}{\partial v} = 0 \tag{3.183}$$

which is given by $K(x, v) = G(C(x, v))$, where $C(x, v)$ is the characteristic curve obtained from the integration of the first two terms of the following equations for the characteristics

$$\frac{dx}{v} = \frac{mdv}{(F - \alpha v^2)(1 - v^2/c^2)^{3/2}} = \frac{dK}{0} \tag{3.184}$$

which is given by

$$C(x, v) = mc^2\left\{\frac{\gamma}{1 - a_\alpha} + \frac{\sqrt{a_\alpha}}{(1 - a_\alpha)^{3/2}}\arctan\left(\frac{1}{\gamma}\sqrt{\frac{a_\alpha}{1 - a_\alpha}}\right)\right\} - Fx \,, \tag{3.185}$$

where a_α has been defined as $a_\alpha = \alpha c^2/F$. Note that this parameter has the range $0 \leq a_\alpha \leq 1$, and is singular for $a_\alpha = 1$. Note also that the characteristic curve has the limit $\lim_{a_\alpha \to 0} C = \gamma mc^2$. Applying the initial data to the solution, it follows that

$$(\gamma - 1)mc^2 = G\left(mc^2\left\{\frac{\gamma}{1-a_\alpha} + \frac{\sqrt{a_\alpha}}{(1-a_\alpha)^{3/2}}\arctan\left(\frac{1}{\gamma}\sqrt{\frac{a_\alpha}{1-a_\alpha}}\right)\right\}\right).$$
(3.186)

So, to find the functionality of G one proceeds as usual

$$\sigma = mc^2\left\{\frac{\gamma}{1-a_\alpha} + \frac{\sqrt{a_\alpha}}{(1-a_\alpha)^{3/2}}\arctan\left(\frac{1}{\gamma}\sqrt{\frac{a_\alpha}{1-a_\alpha}}\right)\right\},$$
(3.187)

and from here it is clear that is not possible to get $\gamma = \gamma(\sigma)$ in general but for small values of $\sqrt{a_\alpha/(1-a_\alpha)}/\gamma$. For this particular weak dissipation case, the above expression is given as

$$\sigma = \frac{mc^2}{(1-a_\alpha)^2}\left\{\gamma(1-a_\alpha) + \frac{a_\alpha}{\gamma}\right\}$$
(3.188)

which has the solution

$$mc^2\gamma_\pm = \frac{(1-a_\alpha)\sigma}{2} \pm \sqrt{\left(\frac{(1-a_\alpha)\sigma}{2}\right)^2 - \frac{a_\alpha m^2 c^4}{1-a_\alpha}}.$$
(3.189)

That is, even in this case the relationship between γ and σ corresponds to two values function. In this way, $G(\sigma)$ is given by

$$G(\sigma) = \frac{(1-a_\alpha)\sigma}{2} \pm \sqrt{\left(\frac{(1-a_\alpha)\sigma}{2}\right)^2 - \frac{a_\alpha m^2 c^4}{1-a_\alpha}} - mc^2,$$
(3.190)

and the constant of motion for this weak dissipation case would be given by

$$K(x,v) = \frac{1-a_\alpha}{2}\left\{mc^2\left[\frac{\gamma}{1-a_\alpha} + \frac{a_\alpha}{\gamma(1-a_\alpha)^2}\right] - Fx\right\}$$

$$\pm\sqrt{\left\{\frac{mc^2}{2}\left[\gamma + \frac{a_\alpha}{\gamma(1-a_\alpha)}\right] - \frac{(1-a_\alpha)Fx}{2}\right\}^2 - \frac{a_\alpha m^2 c^4}{1-a_\alpha}}. \quad (3.191)$$

3.11 Vlasov Equation and Bunched Beam Instabilities

Consider an ensemble of N non interacting particles moving with n-degreed of freedom where each particle obeys the same dynamical system defined by the equation

$$\frac{d\mathbf{x}}{dt} = \mathbf{F}(\mathbf{x},t), \quad \mathbf{x} \in \mathbb{R}^{2n}, \quad t \in \mathbb{R}, \quad \mathbf{F} = (f_1,\ldots,f_{2n}), \quad (3.192)$$

where "t" is the parameter which describes the evolution of the system (non autonomous system). The ensemble of N-particles is described by an scalar function $\psi(\mathbf{x}, t)$ such that

$$\int_{\mathbb{R}^{2n}} \psi(\mathbf{x}, t)d\mathbf{x} = N , \qquad \text{and} \qquad \frac{d\psi}{dt} = 0 , \qquad (3.193)$$

that is, the function is normalized to the number of particles, and it is a constant of motion where its explicit expression is

$$\sum_{i=1}^{2n} f_i(\mathbf{x}, t)\frac{\partial \psi}{\partial x_i} + \frac{\partial \psi}{\partial t} = 0 . \qquad (3.194)$$

There is not really a restriction for an even dimensional dynamical system Eq. (3.192), one will take this case for the applications below. For the particular case when Eq. (3.192) represents a Hamiltonian system, one has that $\mathbf{x} = (\mathbf{q}, \mathbf{p})$ with $\mathbf{q}, \mathbf{p} \in \mathbb{R}^n$ and the dynamical system looks like

$$\frac{d\mathbf{q}}{dt} = \nabla_q H , \qquad \frac{d\mathbf{p}}{dt} = -\nabla_p H , \qquad (3.195)$$

where $H = H(\mathbf{q}, \mathbf{p}, t)$ is the Hamiltonian of the system, we will have the function $\psi = \psi(\mathbf{q}, \mathbf{p}, t)$, and the Vlasov equation will have the form

$$\sum_{j=1}^{n} \left(\frac{\partial H}{\partial p_j}\frac{\partial \psi}{\partial q_j} - \frac{\partial H}{\partial q_j}\frac{\partial \psi}{\partial p_j} \right) + \frac{\partial \psi}{\partial t} = 0 . \qquad (3.196)$$

If the system is autonomous (that is, $\mathbf{F} = \mathbf{F}(\mathbf{x})$, or $H = H(\mathbf{q}, \mathbf{p})$), the term $\partial \psi/\partial t$ can be neglected on the Eq. (3.194) and Eq. (3.196).

3.11.1   *Potential Well Distortion*

Consider now a continuous beam of charged particles (a single bunch) moving inside a beam pipe of a circular accelerator (or stored ring) with a revolution frequency of $\omega_o = 2\pi/T_o$ (T_o is the revolution period) and with an energy E. The relative error in the energy of the beam, $\delta = \Delta E/E$, and the arrival time displacement of the beam at the accelerating cavity, τ, evolve with respect the longitudinal variable "s" according to the following dynamical system

$$\frac{d\tau}{ds} = -\frac{\alpha}{c}\delta , \qquad \frac{d\delta}{ds} = g(\tau) , \qquad (3.197)$$

where α is a positive constant called the momentum compaction factor, c represents the speed of light, and $g(\tau)$ is a function which takes into account the damping and the wake field effect (function $W(\tau)$ generated in a structure of length L) on the beam,

$$g(\tau) = \frac{\omega_s^2}{\alpha c}\tau - \frac{e^2 L}{ET_o c}\int_\tau^\infty h(\tau')W(\tau' - \tau)d\tau' , \qquad (3.198)$$

where "e" is the charge of the particles, ω_s is the synchrotron oscillation frequency ($\omega_s \leq \omega_o$), and $h(\tau)$ is the reduced density part associated to the variable τ,

$$h(\tau) = \int_{\mathbb{R}} \psi(\tau, \delta) \, d\delta . \qquad (3.199)$$

Our system is an autonomous Hamiltonian system with Hamiltonian given by

$$H(\tau, \delta) = \int_0^\tau g(\tau') \, d\tau' + \frac{\alpha}{2c}\delta^2 . \qquad (3.200)$$

Thus, the Vlasov equation ($n = 1$) for an ensemble of charged particles satisfying this dynamical system is given by

$$g(\tau)\frac{\partial\psi}{\partial\delta} - \frac{\alpha\delta}{c}\frac{\partial\psi}{\partial\tau} = 0 . \qquad (3.201)$$

The equations for its characteristic curves are given by

$$\frac{d\delta}{g(\tau)} = \frac{d\tau}{-\alpha\delta/c} = \frac{d\psi}{0} , \qquad (3.202)$$

where the Hamiltonian Eq. (3.200) is a characteristic curve, and the solution of the equation is

$$\psi(\tau, \delta) = G(H(\tau, \delta)) \qquad (3.203)$$

with G being an arbitrary function. Choosing at $\tau = 0$ a Gaussian distribution with standard deviation σ for ψ,

$$\psi(0, \delta) = \frac{1}{\sigma\sqrt{2\pi}}e^{-\delta^2/2\sigma^2} , \qquad (3.204)$$

it follows that the functionality G is determined by

$$G(\xi) = \frac{1}{\sigma\sqrt{2\pi}}e^{-c\xi/\alpha\sigma^2} . \qquad (3.205)$$

Therefore one gets the solution

$$\psi(\tau, \delta) = \frac{1}{\sigma\sqrt{2\pi}}e^{-\delta^2/2\sigma^2}h(\tau) , \qquad (3.206)$$

where the normalization has taken equal to unit, and $h(\tau)$ is the reduced density defined as

$$h(\tau) = e^{-c/\alpha\sigma^2} \int_0^\tau g(\tau')d\tau' . \qquad (3.207)$$

Substituting Eq. (3.198) in this expression, one gets an integral equation for the determination of the reduced density

$$h(\tau) = e^{-\left[\omega_s^2 \tau^2/2\alpha^2\sigma^2 - (e^2 L/\alpha\sigma^2 ET_o) \int_0^\tau d\tau' \int_{\tau'}^\infty h(\lambda)W(\lambda-\tau') \, d\lambda\right]} \qquad (3.208)$$

which is normally solved by numerical methods. The solution brings about the potential well distortion of an initial Gaussian bunch shape.

3.11.2   *Longitudinal Modes*

A more complicated situation is presented for a function g of the form

$$g(\tau, s) = \frac{\omega_s^2}{\alpha c} - \frac{e^2 \omega_o}{T_o Ec} V(\tau, s) , \qquad (3.209)$$

where $V(\tau, s)$ is defined as

$$V(\tau, s) = e^{-i\Omega s/c} \sum_{k=-\infty}^{+\infty} e^{i(k\omega_o+\Omega)\tau} Z(k\omega_o + \Omega)\hat{h}(k\omega_o + \Omega) , \qquad (3.210)$$

being $Z(\omega)$ the longitudinal impedance of the accelerator (Fourier transformation of the wake field), $\hat{h}(\omega)$ is the Fourier transformation of the reduced density, $V(\tau, s)$ is proportional to the voltage exited by the wake field, produced by $h(\tau)$, at the location "s" and seen by a particle at the point "$c\tau$". We have assumed that there is a disturbance having a single frequency Ω in the accelerator, and multi-turn wakes effect has been included. In this case, the dynamical system is written as

$$\frac{d\tau}{ds} = -\frac{\alpha}{c}\delta , \qquad \frac{d\delta}{ds} = g(\tau, s) \qquad (3.211)$$

which is a non autonomous system, and the associated Vlasov equation is given by

$$g(\tau, s)\frac{\partial\psi}{\partial\delta} - \frac{\alpha\delta}{c}\frac{\partial\psi}{\partial\tau} + \frac{\partial\psi}{\partial s} = 0 . \qquad (3.212)$$

Using Eq. (3.209) in this expression, let us write it in the following form

$$\frac{\omega_s^2}{\alpha c}\frac{\partial\psi}{\partial\delta} - \frac{\alpha\delta}{c}\frac{\partial\psi}{\partial\tau} + \frac{\partial\psi}{\partial s} = \frac{e^2\omega_o}{T_o Ec}V(\tau, s)\frac{\partial\psi}{\partial\delta} , \qquad (3.213)$$

and solve this PDEFO through successive approximation method ($\lim_{n \to \infty} \psi_n = \psi$), where the the nth-approximation is given by

$$\frac{\omega_s^2}{\alpha c}\frac{\partial \psi_n}{\partial \delta} - \frac{\alpha \delta}{c}\frac{\partial \psi_n}{\partial \tau} + \frac{\partial \psi_n}{\partial s} = A_{n-1}(\delta, \tau, s) . \qquad (3.214)$$

The term A_{n-1} has been defined as

$$A_{n-1}(\delta, \tau, s) = \frac{e^2 \omega_o}{T_o Ec}V_{n-1}(\tau, s)\frac{\partial \psi_{n-1}}{\partial \delta} , \qquad (3.215)$$

with V_{n-1} the corresponding term with the reduced density $\hat{h}_{n-1}$, and with ψ_o being the solution of the equation

$$\frac{\omega_s^2}{\alpha c}\frac{\partial \psi_o}{\partial \delta} - \frac{\alpha \delta}{c}\frac{\partial \psi_o}{\partial \tau} = 0 . \qquad (3.216)$$

Since the operator associated to Eq. (3.212) satisfies the Lipschitz condition on a compact set of the variables τ, δ, and s, and it there is not singularities of any ψ_n on this set, the succession converge to the solution of the equation (given the initial value). So, let us assume that his happens and solve the problem at first approximation. The solution at order zero was already done, and it is given by Eq. (3.206), in turns, this implies that $\hat{h}_0(\omega)$ is known. Then, the whole term $A_0(\delta, \tau, s)$ is know. However, for the moment let us assume that the term $A_{n-1}(\delta, \tau, s)$ is fully known and solve the general problem. The equations for the characteristic of Eq. (3.214) are given by

$$\frac{d\delta}{\omega_s^2\tau/\alpha c} = \frac{d\tau}{-\alpha \delta/c} = ds = \frac{d\psi_n}{A_{n-1}(\delta, \tau, s)} . \qquad (3.217)$$

The first two terms of this expression brings about the characteristic curve

$$a_1 = \frac{\omega_s^2}{2\alpha c}\tau^2 + \frac{\alpha}{2c}\delta^2 , \qquad (3.218)$$

from which one gets

$$\delta = \pm\sqrt{\frac{2ca_1}{\alpha} - \frac{\omega_s^2}{\alpha^2}\tau^2} . \qquad (3.219)$$

This expression can be substituted in the second term and together with the third term, one obtains the second characteristic curve

$$a_2 = \arcsin\left(\tau\sqrt{\frac{\omega_s^2}{2c\alpha a_1}}\right) \pm \frac{2\alpha a_1}{\omega_s}s. \qquad (3.220)$$

From these two characteristics one gets $\tau = \tau(s)$ and $\delta = \delta(s)$,

$$\tau(s) = \sqrt{\frac{2c\alpha a_1}{\omega_s^2}}\sin\left(\frac{2\alpha a_1}{\omega_s}s + a_2\right) , \quad \delta(s) = \pm\sqrt{\frac{2ca_1}{\alpha}}\cos\left(\frac{2\alpha a_1}{\omega_s}s + a_2\right) . \qquad (3.221)$$

These characteristics are kept independently of the perturbation order. Thus, one can use the last two terms or Eq. (3.217) to get

$$\psi_n(\delta, \tau, s) = \int_0^s A_{n-1}(\delta(s'), \tau(s'), s') \, ds' . \tag{3.222}$$

So, the solution at order zero is given by Eq. (3.206),

$$\psi_0(\delta, \tau) = \frac{1}{\sigma\sqrt{2\pi}} e^{-\delta^2/2\sigma^2} h(\tau) , \tag{3.223}$$

and using Eq. (3.215), at first order approximation the density of particles is given by

$$\psi_1(\delta, \tau, s) = -\frac{e^2\omega_o}{T_o E c \sigma^2} \int_0^s \delta(s') V_0(\tau(s'), s') \psi_0(\delta(s'), \tau(s')) \, ds'. \tag{3.224}$$

3.11.3   *Transverse Modes*

In this case, the beam has a dipole moment in the transversal plane (let us choose the y-direction), and the dynamical equations of motion are given by

$$\frac{d\tau}{ds} = -\frac{\alpha}{c}\delta , \qquad \frac{d\delta}{ds} = \frac{\omega_s^2}{\alpha c}\tau + \frac{y}{cE}\frac{\partial F_y}{\partial \tau} \tag{3.225}$$

and

$$\frac{dy}{ds} = p_y , \qquad \frac{dp_y}{ds} = -\frac{\omega_\beta^2}{c^2}y + \frac{F_y}{E} , \tag{3.226}$$

where $F_y = F_y(\tau, s)$ is the transversal wake force generated by the dipole moment of the beam, ω_β is the unperturbed betatron (transversal) frequency of the beam, and the meaning of other parameters is the same as before. The associated Vlasov equation to the density of charged particles has the following expression

$$\left(\frac{\omega_s^2}{\alpha c}\tau + \frac{y}{cE}\frac{\partial F_y}{\partial \tau}\right)\frac{\partial \psi}{\partial \delta} - \frac{\alpha\delta}{c}\frac{\partial \psi}{\partial \tau} + p_y\frac{\partial \psi}{\partial y} + \left(\frac{F_y}{E} - \frac{\omega_\beta^2}{c^2}y\right)\frac{\partial \psi}{\partial p_y} + \frac{\partial \psi}{\partial s} = 0, \tag{3.227}$$

with $\psi = \psi(\mathbf{x}, s)$, $\mathbf{x} = (\delta, \tau, y, p_y) \in \mathbb{R}^4$, being the density function. As before, let us write this equation in the form

$$\frac{\omega_s^2\tau}{\alpha c}\frac{\partial \psi}{\partial \delta} - \frac{\alpha\delta}{c}\frac{\partial \psi}{\partial \tau} + p_y\frac{\partial \psi}{\partial y} + \left(\frac{F_y}{E} - \frac{\omega_\beta^2}{c^2}y\right)\frac{\partial \psi}{\partial p_y} + \frac{\partial \psi}{\partial s} = -\frac{y}{cE}\left(\frac{\partial F_y}{\partial \tau}\right)\frac{\partial \psi}{\partial \delta}, \tag{3.228}$$

and solve it through successive approximations, $\{\psi_n\}$, such that the nth-approximation satisfies the PDEFO

$$\frac{\omega_s^2 \tau}{\alpha c} \frac{\partial \psi_n}{\partial \delta} - \frac{\alpha \delta}{c} \frac{\partial \psi_n}{\partial \tau} + p_y \frac{\partial \psi_n}{\partial y} + \left(\frac{F_y}{E} - \frac{\omega_\beta^2}{c^2} y \right) \frac{\partial \psi_n}{\partial p_y} + \frac{\partial \psi_n}{\partial s} = A_{n-1}(\mathbf{x}, s),$$

(3.229)

where A_{n-1} has been defined as

$$A_{n-1}(\mathbf{x}, s) = -\frac{y}{cE} \left(\frac{\partial F_y}{\partial \tau} \right) \frac{\partial \psi}{\partial \delta}.$$

(3.230)

The equations for its characteristics are given by

$$\frac{d\delta}{\omega_s^2 \tau / \alpha c} = \frac{d\tau}{-\alpha \delta / c} = \frac{dy}{p_y} = \frac{dp_y}{F_y / E - \omega_\beta^2 y / c^2} = ds = \frac{d\psi_n}{A_{n-1}(\mathbf{x}, s)}.$$

(3.231)

From the first two terms, we can see that we can still have the same characteristics as before,

$$a_1 = \frac{\omega_s^2}{2\alpha c} \tau^2 + \frac{\alpha}{2c} \delta^2$$

(3.232)

and

$$a_2 = \arcsin \left(\tau \sqrt{\frac{\omega_s^2}{2c\alpha a_1}} \right) \pm \left(\frac{2\alpha a_2}{\omega_s} \right) s.$$

(3.233)

Therefore, we can have the same s-dependence for τ and δ,

$$\tau(s) = \sqrt{\frac{2c\alpha a_1}{\omega_s^2}} \sin \left(\frac{2\alpha a_1}{\omega_s} s + a_2 \right)$$

(3.234)

$$\delta(s) = \pm \sqrt{\frac{2c a_1}{\alpha}} \cos \left(\frac{2\alpha a_1}{\omega_s} s + a_2 \right).$$

(3.235)

Now, third, fourth and fifth terms represent Eq. (3.226). So, making the differentiation of y', one gets

$$\frac{d^2 y}{ds^2} + \frac{\omega_\beta^2}{c^2} y = F_y(\tau(s), s) / E$$

(3.236)

which has the solution

$$y(s) = \frac{\omega_\beta}{Ec} \sin \frac{\omega_\beta s}{c} \int \tilde{F}(s) \cos \frac{\omega_\beta s}{c} \, ds + a_3 \sin \frac{\omega_\beta s}{c}$$
$$- \frac{\omega_\beta}{Ec} \cos \frac{\omega_\beta s}{c} \int \tilde{F}(s) \sin \frac{\omega_\beta s}{c} \, ds + a_4 \cos \frac{\omega_\beta s}{c},$$

(3.237)

where the function $\tilde{F}$ has been defined as $\tilde{F}(s) = F_y(\tau(s), s)$. Defining now the functions $f_1(s)$ and $f_2(s)$ as

$$f_1(s) = \frac{\omega_\beta}{Ec} \sin \frac{\omega_\beta s}{c} \int \tilde{F}(s) \cos \frac{\omega_\beta s}{c} \, ds \qquad (3.238)$$

and

$$f_2(s) = \frac{\omega_\beta}{Ec} \cos \frac{\omega_\beta s}{c} \int \tilde{F}(s) \sin \frac{\omega_\beta s}{c} \, ds. \qquad (3.239)$$

Thus, the characteristics a_3 and a_4 have the following form

$$a_3 = \frac{c}{\omega_\beta} \left\{ \frac{\omega_\beta}{c} (y - f_1(s) + f_2(s)) \sin \frac{\omega_\beta s}{c} + (p_y - f_1'(s) - f_2'(s)) \cos \frac{\omega_\beta s}{c} \right\}$$
$$(3.240)$$

and

$$a_4 = \frac{c}{\omega_\beta} \left\{ \frac{\omega_\beta}{c} (y - f_1(s) + f_2(s)) \cos \frac{\omega_\beta s}{c} - (p_y - f_1'(s) - f_2'(s)) \sin \frac{\omega_\beta s}{c} \right\},$$
$$(3.241)$$

where f_1' and f_2' represent the differentiation of these function with resect to "s." In addition, the explicit dependence of y and p_y on the variable "s" has the following form

$$y(s) = f_1(s) - f_2(s) + a_3 \sin \frac{\omega_\beta s}{c} + a_4 \cos \frac{\omega_\beta s}{c} \qquad (3.242)$$

and

$$p_y(s) = f_1'(s) - f_2'(s) + \frac{\omega_\beta}{c} \left[a_3 \cos \frac{\omega_\beta s}{c} - a_4 \sin \frac{\omega_\beta s}{c} \right]. \qquad (3.243)$$

In this way, from Eq. (3.230), one gets the function A_{n-1} fully expressed as a function of the parameter s,

$$\tilde{A}_{n-1}(s) = A_{n-1}(\mathbf{x}(s), s), \qquad (3.244)$$

and the integration of the last two term of Eq. (3.229) can be done,

$$\psi_n(\mathbf{x}, s) = \int_0^s \tilde{A}_{n-1}(s') \, ds', \qquad (3.245)$$

or

$$\psi_n(\mathbf{x}, s) = -\frac{1}{Ec} \int_0^s y(s') \left(\frac{\partial F_y}{\partial \tau} \right)_{\tau(s'), s'} \left(\frac{\partial \psi_{n-1}}{\partial \delta} \right)_{\mathbf{x}(s'), s'} ds'. \qquad (3.246)$$

It is possible to solve the problem at any order of approximation by knowing the zero order density function ψ_0. This zero order approximation satisfies the following PDEFO

$$\frac{\omega_s^2 \tau}{\alpha c}\frac{\partial \psi_0}{\partial \delta} - \frac{\alpha \delta}{c}\frac{\partial \psi_0}{\partial \tau} + p_y\frac{\partial \psi_0}{\partial y} + \left(\frac{F_y}{E} - \frac{\omega_\beta^2}{c^2}y\right)\frac{\partial \psi_0}{\partial p_y} + \frac{\partial \psi_0}{\partial s} = 0. \quad (3.247)$$

Since the equations for its characteristics are almost the same as before,

$$\frac{d\delta}{\omega_s^2 \tau/\alpha c} = \frac{d\tau}{-\alpha \delta/c} = \frac{dy}{p_y} = \frac{dp_y}{F_y/E - \omega_\beta^2 y/c^2} = ds = \frac{d\psi_0}{0} , \quad (3.248)$$

its characteristics are the same (a_1, a_2, a_3, and a_4 above), and the general solution of this equation is given by

$$\psi_0(\mathbf{x}, s) = G(a_1, a_2, a_3, a_4) , \quad (3.249)$$

where G is an arbitrary function of these characteristics. This functionality is determined by the initial distribution of the particles. Suppose that at $s = 0$ one has $\tau(0) = 0$, $p_y(0) = 0$ and $F_y(0,0) = 0$, and $\psi_0(\mathbf{x}(0), 0)$ is having a Gaussian distribution in the variables δ and y with standard deviations σ and σ_y. Then, from the above expression for the characteristics, it follows that

$$\psi_0(\mathbf{x}(0), 0) = G(\alpha \delta^2/2c, 0, y, 0) = \frac{1}{\sigma\sqrt{2\pi}}e^{-\delta^2/2\sigma^2}\frac{1}{\sigma_y\sqrt{2\pi}}e^{-y^2/2\sigma_y^2} . \quad (3.250)$$

Thus, the functionality of G for any variables $(\xi_1, \xi_1, \xi_3, \xi_4)$ is written as

$$G(\xi_1, \xi_2, \xi_3, \xi_4) = \frac{1}{2\pi\sigma\sigma_y}e^{-(c\xi/2\sigma^2 + \xi_3^2/2\sigma_y^2)} , \quad (3.251)$$

and the solution is finally gotten as

$$\psi_0(\mathbf{x}, s) = \frac{1}{2\pi\sigma\sigma_y}e^{-(ca_1/2\sigma^2 + a_3^2/2\sigma_y^2)} , \quad (3.252)$$

where $a_1 = a_1(\delta, \tau)$ is given by Eq. (3.232), and $a_3 = a_3(y, p_y, s)$ is given by Eq. (3.240). Since $\partial\psi_0/\partial\delta = -(\alpha/\sigma^2)\delta\psi_0$, at first order approximation the solution is given by the following integration

$$\psi_1(\mathbf{x}, s) = \frac{\alpha}{Ec\sigma^2}\int_0^s \delta(s')y(s')\psi_0(\mathbf{x}(s'), s')\left(\frac{\partial \tilde{F}_y}{\partial \tau}\right)_{(\mathbf{x}(s'), s')} ds' . \quad (3.253)$$

3.12   Interaction Plasma-Electromagnetic Field

A plasma is a set of heavy charged particles called ions together with very
light charged particles called electrons in some compact region $\Omega \subset \mathbb{R}^3$
such that outside some finite length (Debye length)[1] from this region the
set looks as neutral. Plasma covers a large region of temperatures and
densities, solar wind, solar corona, earth ionosphere, flames, fusion reactors,
laser, plasma, and so on. One could say that there are two ways to treat
the plasma depending on whether or not this one is enough diluted since
this characteristic determine the interaction behavior among the elements
of the plasma.

3.12.1   *Diluted Plasma*

If there is not transport of ions, and the plasma is enough diluted such
that almost not collision occur among ions or electrons, this plasma can be
described by a distribution function $f(\mathbf{x}, \mathbf{v}, t)$ associated to the electrons
which satisfies the Vlasov's equation

$$\mathbf{v} \cdot \frac{\partial f}{\partial \mathbf{x}} + \frac{e}{m} \left[\mathbf{E} + \frac{\mathbf{v}}{c} \cdot \mathbf{B} \right] \cdot \frac{\partial f}{\partial \mathbf{v}} + \frac{\partial f}{\partial t} = 0 \ , \qquad (3.254)$$

where $\mathbf{E}$ and $\mathbf{B}$ are the electric and magnetic fields which are solutions of the
Maxwell's equations (CGS units). For a isotropic-linear-homogeneous, non
dispersed, and local plasma with constant permittivity (ϵ) and permeability
(μ) are given by

$$\nabla \cdot \mathbf{E} = 4\pi\rho/\epsilon \ , \quad \nabla \times \mathbf{E} = -\frac{1}{c}\frac{\partial \mathbf{B}}{\partial t} \ , \quad \nabla \cdot \mathbf{B} = 0 \ , \quad \nabla \times \mathbf{B} = \frac{4\pi\mu}{c}\mathbf{J} + \frac{\epsilon\mu}{c}\frac{\partial \mathbf{E}}{\partial t}$$
$$(3.255)$$

ρ and $\mathbf{J}$ are the charged and current densities which are given in terms of
the function f as

$$\rho(\mathbf{x}, t) = \rho_i - e \int_{\mathbb{R}^3} f(\mathbf{x}, \mathbf{v}, t) d\mathbf{v} \ , \qquad \mathbf{J}(\mathbf{x}, t) = \mathbf{J}_i - e \int_{\mathbb{R}^3} \mathbf{v} f(\mathbf{x}, \mathbf{v}, t) d\mathbf{v}$$
$$(3.256)$$

with ρ_i and $\mathbf{J}_i$ being the densities of charge and current ions, and it is
assumed that an equilibrium situation the distribution of electrons is stationary, depending only on the variable $\mathbf{v}$, and it is of the type Maxwell

---

[1]This Debye length is given by $\lambda_D = \sqrt{k_B T/4\pi e^2 n}$, where k_B is the Boltzmann constant, T is the temperature, e and n are the electron charge and density, which can be
written as $\lambda_D = 740\sqrt{T/n} \ cm$ with T given in eV and n in cm^{-3}, or $\lambda_D = 6.65\sqrt{T/n} \ cm$
when T is given in Kelvin degrees.

distribution function at a temperature T,

$$f_0(\mathbf{v}) = \frac{\rho_i}{e} \sqrt{2(\beta m)^3/\pi} \; v^2 e^{-\beta m v^2/2} \;, \tag{3.257}$$

satisfying

$$e \int_{\mathbb{R}^3} f_0(\mathbf{v})d\mathbf{v} = \rho_i \quad \text{and} \quad e \int_{\mathbb{R}^3} \mathbf{v} f_0(\mathbf{v})d\mathbf{v} = \rho_i \mathbf{v} = \mathbf{J_i} \tag{3.258}$$

because no net charge or current must appear in the plasma ($\beta = 1/k_B T$ with k_B being the Boltzmann constant). The electric and magnetic fields of Eq. (3.255) represent the outside field (homogeneous solution) and the field produced by the charges (particular solutions). Eq. (3.254) represents a non linear integro-partial-differential equation for the function f which is the type of equation outside the topics of this book. However, we note that the electromagnetic field is a linear operator acting on the function f, represented by the solution of the Eqs. (3.255) and (3.256). The solution is in fact the retarded electromagnetic field (see reference J.D. Jackson and reference J.R. Reitz), and let us call this vector operator $\mathbf{F}(\mathbf{x}, \mathbf{v}, t; f)$, and note that for our equilibrium distribution (if there is not external electromagnetic field), one has $\mathbf{F}(\mathbf{x}, \mathbf{v}, t; f_0) = \mathbf{0}$,

$$\mathbf{F}(\mathbf{x}, \mathbf{v}, t; f) = \frac{e}{m} \left[\mathbf{E} + \frac{1}{c} \mathbf{v} \times \mathbf{B} \right] \;. \tag{3.259}$$

In this way, one may propose an iterative method to solve this equation of the form

$$f(\mathbf{x}, \mathbf{v}, t) = \sum_{s=0}^{\infty} f_s(\mathbf{x}, \mathbf{v}, t) \;, \tag{3.260}$$

such that f_0 is our stationary Maxwell's distribution function. Substituting this expression in Eq. (3.254) and separating terms of the same order in f_s, one gets the relation

$$\mathbf{v} \cdot \frac{\partial f_s}{\partial \mathbf{x}} + \frac{\partial f_s}{\partial t} = -\sum_{l=0}^{s} \mathbf{F}(\mathbf{x}, \mathbf{v}, t, f_l) \cdot \frac{\partial f_{s-l}}{\partial \mathbf{v}} \;, \qquad s \geq 1 \;. \tag{3.261}$$

For $s = 1$ and using the above observation, one gets

$$\mathbf{v} \cdot \frac{\partial f_1}{\partial \mathbf{x}} + \frac{\partial f_1}{\partial t} = -\mathbf{F}(\mathbf{x}, \mathbf{v}, t, f_1) \cdot \frac{\partial f_0}{\partial \mathbf{v}} \;. \tag{3.262}$$

For $s = 2$, one has

$$\mathbf{v} \cdot \frac{\partial f_2}{\partial \mathbf{x}} + \frac{\partial f_2}{\partial t} = -\mathbf{F}(\mathbf{x}, \mathbf{v}, t, f_2) \cdot \frac{\partial f_0}{\partial \mathbf{v}} - \mathbf{F}(\mathbf{x}, \mathbf{v}, t, f_1) \cdot \frac{\partial f_1}{\partial \mathbf{v}} \;. \tag{3.263}$$

And so on. These type of equations are linear-integro-partial-differential equations.

If we see Eq. (3.261) as a linear PDEFO in $\mathbb{R}^7$, this equation would have the following equations for their characteristics

$$dt = \frac{dv_1}{0} = \frac{dv_2}{0} = \frac{dv_3}{0} = \frac{dx_1}{v_1} = \frac{dx_2}{v_2} = \frac{dx_3}{v_3} = \frac{f_s}{-\sum_{l=0}^{s} \mathbf{F}\left(\mathbf{x}, \mathbf{v}, t, f_l\right) \cdot \frac{\partial f_{s-l}}{\partial \mathbf{v}}}$$
(3.264)

which would have the characteristics

$$c_i = v_i \ , \quad d_i = x_i - v_i t \quad i = 1, 2, 3 \tag{3.265}$$

independently of s approximation. Now, writing explicitly the dependence on the type of operator, $\int f(\mathbf{x}, \mathbf{v}, t) d\mathbf{v}$, from the first and last terms of Eq. (3.264), and using these characteristics, one has that the density distribution f_s comes from the solution of the following equation

$$\frac{df_s}{dt} + \sum_{l=1}^{s} \mathbf{F}\left(\mathbf{d} + \mathbf{c}t, \mathbf{c}, t; \int_{\mathbb{R}^3} f_l(\mathbf{d} + \mathbf{c}t, \mathbf{v}, t) d\mathbf{v}\right) \cdot \left(\frac{\partial f_{s-l}}{\partial \mathbf{v}}\right)_{(\mathbf{d}+\mathbf{c}t, \mathbf{c}, t)} = 0 \ ,$$
(3.266)

and given the initial conditions $f_j(0)$ for $j = 1, \ldots, s$ the unique solution of this equation can be determined. Since $f_s \geq 0$ for $\mathbf{x}, \mathbf{v} \in \mathbb{R}^3$ and $t \in \mathbb{R}$, the Laplace transform

$$\tilde{f}_s(p) = \mathcal{L}\{f_s(t)\} = \int_0^\infty e^{-pt} f(t) \ dt,$$

being p a complex number, can be used to to solve this equation. Due to $\mathcal{L}$ is a linear operation and has the properties

$\mathcal{L}\{df/dt\} = p\mathcal{L}\{f\} + f(0)$,

$\mathcal{L}\{f \cdot g\} = (1/2\pi i) \int_{Re\ p-i\infty}^{Re\ p+i\infty} \tilde{f}(\sigma)\tilde{g}(p-\sigma)d\sigma$,

$\mathcal{L}\{\int g(t)dt\} = \mathcal{L}\{g\}/p + (1/p)[\int g(t)dt]_{t=0}$, and

$\mathcal{L}\{\int_0^t f(\sigma)g(t-\sigma)d\sigma\} = \mathcal{L}\{f\} \cdot \mathcal{L}\{g\}$.

Applying this transformation in Eq. (3.266), one gets

$$p\tilde{f}_s + f_s(0) + \sum_{l=1}^{s} \widetilde{\mathbf{F}} \star \left(\widetilde{\frac{\partial f_{s-l}}{\partial \mathbf{v}}}\right) = 0 \ , \tag{3.267}$$

Now, using the retarded fields in Eq. (3.259) and proceeding with Eq. (3.266) and Eq. (3.267) is one option to find f_s. However, it is more convenient here to use the representation of the equations in the Fourier space, where the Fourier transformation

$$\hat{f}(\mathbf{k}) = \mathcal{F}[f(\mathbf{x})] = \frac{1}{(2\pi)^{3/2}} \int_{\mathbb{R}^3} e^{i\mathbf{x}\cdot\mathbf{k}} f(\mathbf{x})\, d\mathbf{x} \qquad (3.268)$$

is a linear operation having the property $\mathcal{F}[\partial f/\partial x_n] = -ik_n \mathcal{F}[f]$. Using this property in Eq. (3.255), it follows that

$$ik\cdot\widehat{\mathbf{E}} = 4\pi\hat{\rho}/\epsilon, \quad ik\times\widehat{\mathbf{E}} = -\frac{1}{c}\frac{\partial\widehat{\mathbf{B}}}{\partial t}, \quad ik\cdot\widehat{\mathbf{B}} = 0, \quad ik\times\widehat{\mathbf{B}} = \frac{4\pi\mu}{c}\hat{\mathbf{J}} + \frac{\epsilon\mu}{c}\frac{\partial\widehat{\mathbf{E}}}{\partial t}.$$
$$(3.269)$$

Using the vector identity $\mathbf{a}\times(\mathbf{b}\times\mathbf{c}) = (\mathbf{a}\cdot\mathbf{c})\mathbf{b} - (\mathbf{a}\cdot\mathbf{b})\mathbf{c}$ in the second and last terms and making some rearrangements, the following evolution equation are obtained in the Fourier space

$$\frac{\partial^2\widehat{\mathbf{E}}}{\partial t^2} + \omega^2\widehat{\mathbf{E}} = -i\frac{4\pi c\hat{\rho}}{\epsilon\mu}\,\mathbf{k} - \frac{4\pi}{\epsilon}\frac{\partial\hat{\mathbf{J}}}{\partial t} \qquad (3.270)$$

$$\frac{\partial^2\widehat{\mathbf{B}}}{\partial t^2} + \omega^2\widehat{\mathbf{B}} = i\frac{4\pi c}{\epsilon}\,\mathbf{k}\times\hat{\mathbf{J}} \qquad (3.271)$$

where ω has been defined as $\omega = ck/\sqrt{\epsilon\mu}$. Ignoring the homogeneous solutions of these equations, their particular solutions are given by

$$\widehat{\mathbf{E}}_p = -\frac{i4\pi c}{\epsilon\mu\omega}\mathbf{k}\int_0^t \hat{\rho}(s)\sin\omega(t-s)ds - \frac{4\pi}{\epsilon\omega}\int_0^t \frac{\partial\hat{\mathbf{J}}(s)}{\partial s}\sin\omega(t-s)ds \,(3.272)$$

$$\widehat{\mathbf{B}}_p = \frac{i4\pi c}{\epsilon\omega}\mathbf{k}\times\int_0^t \hat{\mathbf{J}}(s)\sin\omega(t-s)ds \,, \qquad (3.273)$$

where the dependence only on time for the current and density has been written for short writing. The Fourier transformation of the continuity equation $\nabla\cdot\mathbf{J} + \partial\rho/\partial t = 0$ brings about the relation

$$i\mathbf{k}\cdot\hat{\mathbf{J}} + \frac{\partial\hat{\rho}}{\partial t} = 0 \,. \qquad (3.274)$$

Let us solve Eq. (3.262) using these tools and noting that $\partial f_0/\partial\mathbf{v}$ depends only on $\mathbf{v}$. Applying the Fourier transformation to this equation, one gets

$$i\mathbf{v}\cdot\mathbf{k}\hat{f}_1 + \frac{\partial\hat{f}_1}{\partial t} = -\frac{e}{m}\left[\widehat{\mathbf{E}} + \frac{1}{c}\mathbf{v}\times\widehat{\mathbf{B}}\right]\cdot\frac{\partial f_0}{\partial\mathbf{v}} \qquad (3.275)$$

or

$$iv \cdot \mathbf{k} \, \hat{f}_1 + \frac{\partial \hat{f}_1}{\partial t} = \frac{i4\pi ce}{\epsilon\mu\omega m} \mathbf{k} \cdot \frac{\partial f_0}{\partial \mathbf{v}} \int_0^t \hat{\rho}(s) \sin\omega(t-s)ds$$

$$+ \left(\frac{4\pi e}{\epsilon\omega m} \right) \frac{\partial f_0}{\partial \mathbf{v}} \cdot \int_0^t \frac{\partial \hat{\mathbf{J}}(s)}{\partial s} \sin\omega(t-s)ds$$

$$- \frac{i4\pi ce}{\epsilon\omega m} \mathbf{v} \times \left[\mathbf{k} \times \int_0^t \hat{\mathbf{J}}(s) \sin\omega(t-s)ds \right] \cdot \frac{\partial f_0}{\partial \mathbf{v}}. \quad (3.276)$$

Using the same above vector identity, this equation can be written as

$$iv \cdot \mathbf{k} \, \hat{f}_1 + \frac{\partial \hat{f}_1}{\partial t} = \frac{i4\pi ce}{\epsilon\mu\omega m} \mathbf{k} \cdot \frac{\partial f_0}{\partial \mathbf{v}} \int_0^t \hat{\rho}(s) \sin\omega(t-s)ds$$

$$+ \left(\frac{4\pi e}{\epsilon\omega m} \right) \frac{\partial f_0}{\partial \mathbf{v}} \cdot \int_0^t \frac{\partial \hat{\mathbf{J}}(s)}{\partial s} \sin\omega(t-s)ds$$

$$- \frac{i4\pi ce}{\epsilon\omega m} \left(\mathbf{k} \cdot \frac{\partial f_0}{\partial \mathbf{v}} \right) \left[\mathbf{v} \cdot \int_0^t \hat{\mathbf{J}}(s) \sin\omega(t-s)ds \right]$$

$$+ \frac{i4\pi ce}{\epsilon\omega m} (\mathbf{v} \cdot \mathbf{k}) \left[\frac{\partial f_0}{\partial \mathbf{v}} \cdot \int_0^t \hat{\mathbf{J}}(s) \sin\omega(t-s)ds \right].$$

$$(3.277)$$

Note that for the particular case when one considers $\hat{\mathbf{J}} = \mathbf{0}$, which is equivalent to ignore the magnetic field contribution, the density $\hat{\rho}$ must not depend explicitly on time due to continuity equation Eq. (3.292), and the above equation is reduced to the equation

$$i\mathbf{k} \cdot \mathbf{v} \, \hat{f}_1 + \frac{\partial \hat{f}_1}{\partial t} = \frac{i4\pi ce}{\epsilon\mu\omega^2 m} \mathbf{v} \cdot \left(\frac{\partial f_0}{\partial \mathbf{v}} \right) \hat{\rho}. \quad (3.278)$$

However from Eq. (3.256), one has an explicitly time dependence in the density through the function f_1,

$$\hat{\rho}(\mathbf{k}, t) = -e \int_{\mathbb{R}^3} \hat{f}_1(\mathbf{k}, \mathbf{v}, t)dv. \quad (3.279)$$

Therefore, to solve consistently Eq. (3.276), one needs to take into account the dependence on time of the density. Ignoring this detail, the solution of Eq. (3.278) leads to what is called *"Landau damping effect"* (see reference L. Landau and reference V. Maslo and M. Fedoryuk). With the above properties of the Laplace transformation and knowing that $\mathcal{L}\{\sin\omega t\} = \omega/(p^2 + \omega^2)$, let us apply this transformation to Eq. (3.277) to get the following expression

$$(i\mathbf{v} \cdot \mathbf{k} + p)\tilde{f}_1 = \hat{f}_1(0) + \frac{i4\pi ce}{\epsilon\mu\omega m}\left(\mathbf{k} \cdot \frac{\partial f_0}{\partial \mathbf{v}}\right)\frac{\tilde{\rho}\,\omega}{p^2 + \omega^2}$$

$$+ \left(\frac{4\pi e}{\epsilon\omega m}\right)\frac{\partial f_0}{\partial \mathbf{v}} \cdot \left[\frac{(p\tilde{\mathbf{J}} - \hat{\mathbf{J}}(0))\,\omega}{p^2 + \omega^2}\right]$$

$$- \frac{i4\pi ce}{\epsilon\omega m}\left(\mathbf{k} \cdot \frac{\partial f_0}{\partial \mathbf{v}}\right)\left[\frac{\mathbf{v} \cdot \tilde{\mathbf{J}}\,\omega}{p^2 + \omega^2}\right]$$

$$+ \frac{i4\pi ce}{\epsilon\omega m}(\mathbf{v} \cdot \mathbf{k})\left[\frac{\frac{\partial f_0}{\partial \mathbf{v}} \cdot \tilde{\mathbf{J}}\,\omega}{p^2 + \omega^2}\right] \quad (3.280)$$

where $\tilde{\rho} = \tilde{\rho}(\mathbf{k}, \mathbf{v}, p)$ and $\tilde{\mathbf{J}} = \tilde{\mathbf{J}}(\mathbf{k}, \mathbf{v}, p)$ are the corresponding expressions in the Fourier-Laplace space. Rearranging terms on this expression, one gets the solution

$$\tilde{f}_1(\mathbf{k}, \mathbf{v}, p) = -\frac{i\hat{f}_1(0)}{(\mathbf{v} \cdot \mathbf{k} - ip)} + \frac{4\pi ce}{\epsilon\mu m(p^2 + \omega^2)(\mathbf{v} \cdot \mathbf{k} - ip)}$$

$$\times \left\{\left(\mathbf{k} \cdot \frac{\partial f_0}{\partial \mathbf{v}}\right)\tilde{\rho} - \frac{i\mu}{c}\frac{\partial f_0}{\partial \mathbf{v}} \cdot (p\tilde{\mathbf{J}} - \hat{\mathbf{J}}(0)) - \mu\left(\mathbf{k} \cdot \frac{\partial f_0}{\partial \mathbf{v}}\right)(\mathbf{v} \cdot \tilde{\mathbf{J}})\right.$$

$$\left. + \mu(\mathbf{v} \cdot \mathbf{k})\left(\frac{\partial f_0}{\partial \mathbf{v}}\right)\right\}. \quad (3.281)$$

The dispersion relation and other physical properties can be obtained by substituting this expression into the definitions

$$\tilde{\rho} = e \int_{\mathbb{R}^3} \tilde{f}_1(\mathbf{k}, \mathbf{v}, p)d\mathbf{v} \quad \text{and} \quad \tilde{\mathbf{J}} = e \int_{\mathbb{R}^3} \mathbf{v}\tilde{f}_1(\mathbf{k}, \mathbf{v}, p)d\mathbf{v}. \quad (3.282)$$

3.12.2 *Diluted Linear, Dispersed and Nonlocal Plasma*

In this case, Maxwell's equations are given by

$$\nabla \cdot \mathbf{D} = 4\pi\rho, \quad \nabla \times \mathbf{E} = -\frac{1}{c}\frac{\partial \mathbf{B}}{\partial t}, \quad \nabla \cdot \mathbf{B} = 0, \quad \nabla \times \mathbf{H} = \frac{4\pi}{c}\mathbf{J} + \frac{1}{c}\frac{\partial \mathbf{D}}{\partial t}$$
$$(3.283)$$

where the relations between the fields $\mathbf{E}, \mathbf{B}$ and $\mathbf{D}, \mathbf{H}$ are given by $\mathbf{D} = \epsilon \star \mathbf{E}$ and $\mathbf{B} = \mu \star \mathbf{H}$, being ϵ and μ the permittivity and permeability functions, and the star product is just the space-time convolution, $(f \star g)(\mathbf{x}, t) = \int f(\mathbf{x} - \mathbf{x}', t - t')g(\mathbf{x}', t')d\mathbf{x}'dt'$, which characterizes the linear, dispersed and nonlocal behavior of the system. We can use the Fourier-Laplace transformation on these equations,

$$\tilde{f}(\mathbf{k}, p) = \mathcal{F}_{\mathcal{L}}[f(\mathbf{x}, t)] = \frac{1}{(2\pi)^{3/2}}\int_{\mathbb{R}^3}\int_0^\infty e^{i\mathbf{k}\cdot\mathbf{x} - pt}f(\mathbf{x}, t)d\mathbf{x}dt, \quad (3.284)$$

having the properties

i) $\mathcal{F}_{\mathcal{L}}[f_1 + \alpha f_2] = \mathcal{F}_{\mathcal{L}}[f_1] + \alpha \mathcal{F}_{\mathcal{L}}[f_2]$

ii) $\mathcal{F}_{\mathcal{L}}[\partial f/\partial x_l] = ik_l \tilde{f}(\mathbf{k}, p)$

iii) $\mathcal{F}_{\mathcal{L}}[\partial f/\partial t] = p\tilde{f}(\mathbf{k}, p) - \hat{f}(\mathbf{k}, 0)$

iv) $\mathcal{F}_{\mathcal{L}}[f \star g] = \tilde{f} \cdot \tilde{g}$,

with $\hat{f}$ being just the Fourier transformation of f, and p is in general a complex number. These Maxwell equation are transformed as

$$i\mathbf{k} \cdot \widetilde{\mathbf{E}} = 4\pi\tilde{\rho}/\tilde{\epsilon} , \qquad i\mathbf{k} \times \widetilde{\mathbf{E}} = -\frac{1}{c}(p\widetilde{\mathbf{B}} - \widehat{\mathbf{B}}_0) \tag{3.285}$$

and

$$i\mathbf{k} \cdot \widetilde{\mathbf{B}} = 0 , \qquad i\mathbf{k} \times \widetilde{\mathbf{B}} = \frac{4\pi\tilde{\mu}}{c}\tilde{\mathbf{J}} + \frac{\tilde{\mu}\tilde{\epsilon}}{c}(p\widetilde{\mathbf{E}} - \widehat{\mathbf{E}}_0) \tag{3.286}$$

where we have defined $\widehat{\mathbf{E}}_0 = \hat{\mathbf{E}}(\mathbf{k}, 0)$ in the Fourier space, similarly with $\widehat{\mathbf{B}}_0$. From Eq. (3.272) and Eq. (3.273), one has that $\widehat{\mathbf{E}}_0 = \widehat{\mathbf{B}}_0 = \mathbf{0}$. Now, using the same vector identity $\mathbf{a} \times (\mathbf{b} \times \mathbf{c}) = (\mathbf{a} \cdot \mathbf{c})\mathbf{b} - (\mathbf{a} \cdot \mathbf{b})\mathbf{c}$ we used before, it follows that

$$\widetilde{\mathbf{E}}(\mathbf{k}, p) = \frac{-1}{k^2 + \frac{\tilde{\mu}\tilde{\epsilon}p^2}{c^2}}\left(i\frac{4\pi\tilde{\rho}}{\tilde{\epsilon}}\mathbf{k} + \frac{4\pi\tilde{\mu}p}{c^2}\tilde{\mathbf{J}}\right) \tag{3.287}$$

$$\widetilde{\mathbf{B}}(\mathbf{k}, p) = i\frac{4\pi\tilde{\mu}/c}{k^2 + \frac{\tilde{\mu}\tilde{\epsilon}p^2}{c^2}} \mathbf{k} \times \tilde{\mathbf{J}}, \tag{3.288}$$

where $\tilde{\rho}$ and $\tilde{\mathbf{J}}$ are, of course, given in terms of the transformed distribution function $\tilde{f}$ as

$$\tilde{\rho}(\mathbf{k}, p) = \rho_i - e\int_{\mathbb{R}^3}\tilde{f}(\mathbf{k}, \mathbf{v}, p)d\mathbf{v} , \qquad \tilde{\mathbf{J}}(\mathbf{k}, p) = \mathbf{J}_i - e\int_{\mathbb{R}^3}\mathbf{v}\tilde{f}(\mathbf{k}, \mathbf{v}, p)d\mathbf{v} . \tag{3.289}$$

At first order approximation, we can calculate the distribution function by taking the Fourier-Laplace transformation to Eq. (3.265), which brigs about the expression

$$\tilde{f}_1(\mathbf{k}, \mathbf{v}, p) = \frac{\hat{f}_1(\mathbf{k}, \mathbf{v}, 0)}{i\mathbf{v} \cdot \mathbf{k} + p} - \frac{1}{i\mathbf{v} \cdot \mathbf{k} + p}\widetilde{\mathbf{F}}(\mathbf{k}, \mathbf{v}, p, \tilde{f}_1) \cdot \frac{\partial f_0}{\partial \mathbf{v}}, \tag{3.290}$$

where $\widetilde{\mathbf{F}}$ is given by

$$\widetilde{\mathbf{F}}(\mathbf{k}, \mathbf{v}, p, \tilde{f}_1) = \frac{e}{m}\left[\widetilde{\mathbf{E}} + \frac{1}{c} \times \widetilde{\mathbf{B}}\right] \tag{3.291}$$

where $\widetilde{\mathbf{E}}$ and $\widetilde{\mathbf{B}}$ are given by Eq. (3.287) and Eq. (3.288). By substituting one expresion into another would appears a relation which brings about the dispersion relation of the system.

3.12.3 Non Dilute Plasma

In this case there are interactions among the different species of ions and electrons which the most probable is the two bodies interaction (ion-ion, ion-electron, or electron-electron). However, it is assume that the probability of having a particular species in a element of phase-space-time $d\mathbf{x}d\mathbf{v}dt$ is independent for each specie. In this way, the distribution function of the species α, $f_\alpha(\mathbf{x}, \mathbf{v}, t)$, having a density

$$\rho_\alpha(\mathbf{x}, t) = \int_{\mathbb{R}^3} f_\alpha(\mathbf{x}, \mathbf{v}, t)d\mathbf{v} , \qquad (3.292)$$

satisfies the so called Boltzmann's equation (under interaction with electromagnetic field)

$$\mathbf{v} \cdot \frac{\partial f_\alpha}{\partial \mathbf{x}} + \frac{e}{m}\left[\mathbf{E} + \frac{\mathbf{v}}{c} \cdot \mathbf{B}\right] \cdot \frac{\partial f_\alpha}{\partial \mathbf{v}} + \frac{\partial f}{\partial t_\alpha} = \left(\frac{\partial f_\alpha}{\partial t}\right)_c , \qquad (3.293)$$

where the term on the right hand side denotes the rate of change of f_α due to collisions. Due to the effect of the screening (Debye length), the Coulomb interaction last for a time $\tau \sim \lambda_D/v$ (being v the relative velocity of ions in collision), and it is assumed that during this time the density distributions of the ions in collision do not change significantly, with a maximum parameter of impact given by the Debye length. There are several approximations for this collision term (see reference J. Sheffield and reference S. Chandrasekhar, and reference C. Cercignani), and we limited ourselves to the case when this term depend only of the same distribution function f_α (for example in the Krook model $(\partial f_\alpha/\partial t)_c = -\nu_{\alpha\beta}(f_\alpha - f_{\alpha 0})$ with $\nu_{\alpha\beta}$ being the collision frequency between the species α and β, and $f_{\alpha 0}$ is the relaxed distribution function). Therefore, dropping the index α in Eq. (3.294), the Boltzmann's equation looks as

$$\mathbf{v} \cdot \frac{\partial f}{\partial \mathbf{x}} + \frac{e}{m}\left[\mathbf{E} + \frac{\mathbf{v}}{c} \cdot \mathbf{B}\right] \cdot \frac{\partial f}{\partial \mathbf{v}} + \frac{\partial f}{\partial t} = C(f) , \qquad (3.294)$$

Consider an external electromagnetic field given by the following expressions

$$\mathbf{E} = \left(E_x(t), E_y(t), E_z(t)\right) , \quad \text{and} \quad \mathbf{B} = \left(0, 0, B\right) , \qquad (3.295)$$

where B is constant. Then, substituting these fields in Eq. (3.294), one gets the following equation

$$\frac{\partial f}{\partial t} + v_x\frac{\partial f}{\partial x} + v_y\frac{\partial f}{\partial y} + v_z\frac{\partial f}{\partial z} + \frac{e}{m}\left(E_x + \frac{v_y B}{c}\right)\frac{\partial f}{\partial v_x}$$
$$+ \frac{e}{m}\left(E_y - \frac{v_x B}{c}\right) + \frac{eE_z}{m}\frac{\partial f}{\partial v_z} = C(f) . \qquad (3.296)$$

This equation represents a quasi-linear PDEFO defined in $\mathbb{R}^7$, and the equations for its characteristics are given by (in parametric form)

$$\frac{dx}{dt} = v_x, \quad \frac{dy}{dt} = v_y, \quad \frac{dz}{dt} = v_z, \quad \frac{dv_x}{dt} = \frac{e}{m}\left(E_x + \frac{v_y B}{c}\right)$$

$$\frac{dv_y}{dt} = \frac{e}{m}\left(E_y - \frac{v_x B}{c}\right), \quad \frac{dv_z}{dt} = \frac{e}{m}E_z, \quad \frac{df}{dt} = C(f). \qquad (3.297)$$

Differentiating the fourth and fifth equations and using them again, one gets the decoupled equations

$$\frac{d^2 v_x}{dt^2} + \omega^2 v_x = \frac{e}{m}\left(\dot{E}_x + E_y\right), \quad \text{and} \quad \frac{d^2 v_y}{dt^2} + \omega^2 v_y = \frac{e}{m}\left(\dot{E}_y - E_x\right), \tag{3.298}$$

where $\dot{\xi}$ means $d\xi/dt$. Since the right hand side of these equation are functions depending on time, the solutions of these equation are given by

$$v_x = c_1 \cos\omega t + c_2 \sin\omega t + \frac{1}{\omega}\int^t g_1(s)\sin\omega(t - s)\, ds \tag{3.299}$$

and

$$v_y = c_3 \cos\omega t + c_4 \sin\omega t + \frac{1}{\omega}\int^t g_2(s)\sin\omega(t - s)\, ds, \tag{3.300}$$

where the functions g_1 and g_2 have been defined as

$$g_1(t) = \frac{e}{m}\left(\dot{E}_x + E_y\right) \quad \text{and} \quad g_2(t) = \frac{e}{m}\left(\dot{E}_y - E_x\right). \tag{3.301}$$

Thus, the first two equations of (3.297) are readily integrated as

$$x = \frac{c_1}{\omega}\sin\omega t - \frac{c_2}{\omega}\cos\omega t + \frac{1}{\omega}\int dt \int^t g_1(s)\sin\omega(t - s)\, ds \tag{3.302}$$

and

$$y = \frac{c_3}{\omega}\sin\omega t - \frac{c_4}{\omega}\cos\omega t + \frac{1}{\omega}\int dt \int^t g_1(s)\sin\omega(t - s)\, ds. \tag{3.303}$$

The integration over the z-dependence is straightforward and is given by

$$v_z = \frac{e}{m}\int^t E_z(s)\, ds + c_5 \quad \text{and} \quad z = \frac{e}{m}\int dt \int^t E_z(s)\, ds + c_5 t + c_6. \tag{3.304}$$

Therefore, the characteristics are given by

$$c_1 = v_x + x\omega + \cos\omega t \int^t g_1(s)\sin\omega(t-s)ds$$

$$+ \sin\omega t \int dt \int^t g_1(s)\sin\omega(t-s)ds \ ,$$

$$c_2 = v_x - x\omega + \sin\omega t \int^t g_1(s)\sin\omega(t-s)ds$$

$$- \cos\omega t \int dt \int^t g_1(s)\sin\omega(t-s)ds \ ,$$

$$c_3 = v_y + y\omega + \cos\omega t \int^t g_2(s)\sin\omega(t-s)ds$$

$$+ \sin\omega t \int dt \int^t g_2(s)\sin\omega(t-s)ds \ ,$$

$$c_4 = v_y - y\omega + \sin\omega t \int^t g_2(s)\sin\omega(t-s)ds$$

$$- \cos\omega t \int dt \int^t g_2(s)\sin\omega(t-s)ds \ ,$$

$$c_5 = v_z - \frac{e}{m} \int^t E_z(s)ds \ ,$$

$$c_6 = z - \frac{e}{m} \int dt \int^t E_z(s)ds - v_z t + \frac{et}{m} \int^t E_z(s)ds \ . \qquad (3.305)$$

Finally, the integration of the last equation in (3.297) can be expressed as

$$\int \frac{df}{C(f)} = t + g(\mathbf{c}(\mathbf{x}, \mathbf{v}, t)) \ , \qquad (3.306)$$

where the vector $\mathbf{c}$ has been defined as $\mathbf{c} = (c_1, c_2, c_3, c_4, c_5, c_6)$, and g is an arbitrary function. For the Krook model we can write the general solution as

$$f(\mathbf{x}, \mathbf{v}, t) = f_{\alpha 0} + G(\mathbf{c}(\mathbf{x}, \mathbf{v}, t))e^{-\nu_{\alpha\beta} t} \ , \qquad (3.307)$$

where G is an arbitrary function which is determined by initial conditions.

3.13 Decoherence in a Quantum System

In the Schrödinger picture of the quantum mechanics one can solve a problem either with the Schrödinger equation for the wave function $|\Psi\rangle$,

$$i\hbar\frac{\partial|\Psi\rangle}{\partial t} = \widehat{H}|\Psi\rangle \ , \qquad (3.308)$$

where $|\Psi\rangle$ is a ket vector element of a finite or infinite Hilbert space (see reference A. Messiah and reference P.A.M. Dirac), and $\widehat{H}$ is an Hermitian operator associated to the Hamiltonian of the system, or we can solve a problem with the von Neuman equation for the density matrix operator ρ,

$$i\hbar\frac{\partial\rho}{\partial t} = [\widehat{H}, \rho] \,, \tag{3.309}$$

where $\rho = |\Psi\rangle\langle\Psi|$ is an Hermitian operator with $\rho \geq 0$, $tr(\rho) = 1$, and $tr(\rho^2) = 1$ (for pure state, $\rho^2 = \rho$, which is solution of (3.308)), $[a, b] = ab - ba$ is the commutator of two operators. If the relation $\rho^2 \neq \rho$ is satisfied, one says the density matrix corresponds to a mixed state. The evolution of the system is unitary in both cases, that is, there is an unitary operator U ($U^\dagger U = I$) such that teh evolution of the systems is given $|\Psi(t)\rangle = U(t)|\Psi_0\rangle$ or $\rho(t) = U(t)\rho_0 U^\dagger(t)$. Although must of the quantum system (QS) are considered insulated from environment, that is not generally true in practice, and the interaction of the quantum system with the environment brings about the relaxation and decoherence of the system. One says that within a quantum system has appeared decoherence when the non diagonal elements of the density matrix become zero (the matrix becomes diagonal). The study of a QS with its environment is called *open quantum system* (OQS), and the Hamiltonian of a OQS would be of the form $\widehat{H} = \widehat{H}_s + \widehat{H}_e + \widehat{H}_{se}$, where $\widehat{H}_{se}$ contains the interaction between system and environment. When this Hamiltonian is introduced in (3.308) and the trace over the environment variables is taken, the resulting density matrix is called *"reduced density matrix,"* and its evolution is not any longer unitary. The resulting evolution equation is called *"master equation of motion,"* and it used to be of Limdblad type of equation (see reference G. Limdblad, and see reference H.P. Breuer and F. Petruccione),

$$i\hbar\frac{\partial\rho}{\partial t} = [\widehat{H}_s, \rho] + \mathcal{L}(\rho) \,, \tag{3.310}$$

where ρ is the reduced density matrix, and $\mathcal{L}$ is a linear operator which contains the terms resulting from the environment and system-environment interaction at some approximation. For a 1-D, QS with continuos variables interacting with the environment at room temperature in a Markovian process (without memory), the master equation is written as (see A.O. Caldeira and A.J. Leggett),

$$i\hbar\frac{\partial\rho}{\partial t} = [\widehat{H}_s, \rho] + \frac{\beta}{2}[\{p, z\}, \rho] - \frac{i}{\hbar}D\beta[z, [z, p]] + 2\beta([z, \rho p] - [p, \rho z]) \,, \tag{3.311}$$

where $\{a, b\} = ab + ba$ is the anti-commutator of two operators, z and p are the position and momentum of the particle, β represents the dissipation coefficient, and $D = 2k_B T/\hbar$ represents the diffusion coefficient which depends on the temperature T.

For a Josephson[2] junction device coupled to LC oscillator via a capacitor, C_m which is coupled to a transmission line via another capacitor, C_c, and with a oscillator capacitance $C_o = C + C_m$, a Hamilton for this quantum system (ignoring the transmission line charge) can be written as (see reference G. Johanson et al, see also reference Y. Makchlin et al)

$$\widehat{H}_S = \frac{1}{2}\left(p^2 + z^2\right) - \frac{\epsilon}{2}\sigma_z - \eta p^2 \sigma_z , \qquad (3.312)$$

where p is proportional to the charge of the magnetic flux oscillator, $p = q_c\left(L/C_o\hbar^2\right)^{1/4}$, the variable z is proportional to this magnetic flux, $z = \phi_c\left(C_o/L\hbar^2\right)^{1/4}$, which are both conjugated, $[z, p] = i$. The evolution is given in terms of the LC value of the oscillator, $\tau = t/\sqrt{LC_o}$, and the Hamiltonian has been written in a normalized way, $\widehat{H}_S = \sqrt{LC_o}\widehat{H}/\hbar$. The parameter $\eta = g_c C_o/2$ measures the coupling between the oscillator and the two states of the superconductor junction (current flowing in one direction or the its opposed direction, like the two states spin one half particle), and $\epsilon = E_J\sqrt{LC_o}/\hbar$ is the normalized energy in the junction (qubit). The reduced density matrix will depend on two continuous variables, z and z', and two discrete variables, s and s', $\rho_{ss'}(z, z') = \langle z, s|\rho|z', s'\rangle$. Using (3.312) in (3.311), the resulting master equation (after some rearrangements) is given by

$$\frac{\partial \rho_{ss'}}{\partial \tau} = \left\{ \frac{i}{2}(\partial_{zz} - \partial_{z'z'}) - \frac{i}{2}(z^2 - z'^2) - i\eta(s\partial_{zz} - s'\partial_{z'z'}) \right.$$

$$\left. - \frac{\beta}{2}(z - z')(\partial_z - \partial_{z'}) - D\beta(z - z')^2 \right\}\rho_{ss'} , \qquad (3.313)$$

where $s, s' = \pm 1/2$. This equation represents a partial differential equation of second order. However, let us make first the following change of variables (see reference G. López and P. López)

$$r = z - z' , \quad R = (z + z')/2, \qquad (3.314)$$

getting the following expression for this equation as

$$\frac{\partial \rho_{ss'}}{\partial \tau} = \left\{ i\partial_{rR} - irR - \beta r\partial_r - i\eta\left[(s - s')\partial_{rr} + (s + s')\partial_{rR} + \frac{s - s'}{4}\partial_{RR}\right] \right.$$

$$\left. - \beta r\partial_r - D\beta r^2 \right\}\rho_{ss'} . \qquad (3.315)$$

---

[2] I want to thank G.P. Berman for encouraging me to look at this problem.

Now, taking the Fourier transformation with respect the variable R, $\hat{\rho}_{ss'}(r, k, \tau) = \int_{\mathbb{R}} e^{ikR} \rho_{ss'}(r, R, \tau) dR$, the following equation is gotten

$$\frac{\partial \hat{\rho}_{ss'}}{\partial \tau} = [k - \beta r - k\eta(s + s')] \frac{\partial \hat{\rho}_{ss'}}{\partial r} - r \frac{\partial \hat{\rho}_{ss'}}{\partial k} - i\eta(s - s') \frac{\partial^2 \hat{\rho}_{ss'}}{\partial r^2}$$

$$- [D\beta r^2 + \frac{s - s'}{4} k^2] \hat{\rho}_{ss'}. \tag{3.316}$$

With respect the current flowing in the superconducting joint (characterized by the discrete variables s and s'), this equation is divided in the diagonal matrix elements $(s = s')$, which satisfies the equation

$$\frac{\partial \hat{\rho}_s}{\partial \tau} + [\beta r - k + 2k\eta s] \frac{\partial \hat{\rho}_s}{\partial r} - r \frac{\partial \hat{\rho}_s}{\partial k} = -D\beta r^2 \hat{\rho}_s, \tag{3.317}$$

and the non diagonal matrix elements $(s = -s')$, which satisfies the equation

$$\frac{\partial \hat{\rho}_s'}{\partial \tau} + [\beta r - k] \frac{\partial \hat{\rho}_s'}{\partial r} - r \frac{\partial \hat{\rho}_s'}{\partial k} = -[D\beta r^2 + \frac{sk^2}{2}] \hat{\rho}_s' - i2s\eta \frac{\partial \hat{\rho}_s'}{\partial r^2}. \tag{3.318}$$

Diagonal elements: Let us deal with Eq. (3.317) which represents a linear PDEFO defined in $\mathbb{R}^3$. The equations for its characteristics in parametric form are given by

$$\frac{dk}{d\tau} = r \ , \quad \frac{dr}{d\tau} = \beta r - k + 2k\eta s \ , \quad \frac{d\rho_s}{d\tau} = -D\beta r^2 \rho_s. \tag{3.319}$$

Making the differentiation of the first term and using the second one, one gets

$$\frac{d^2 k}{d\tau^2} + (1 - 2\eta s)k - \beta \frac{dk}{d\tau} = 0 \tag{3.320}$$

which has the solution

$$k(\tau) = e^{\beta \tau / 2} \left(a_1 \sin \omega \tau + a_2 \cos \omega \tau \right) , \tag{3.321}$$

where ω has been defined as

$$\omega = \sqrt{1 - \beta^2 / 4 - 2\eta s} \ . \tag{3.322}$$

Therefore, using the first term of (3.319), we also have

$$r(\tau) = e^{\beta \tau / 2} \left[a_1 (\omega \cos \omega \tau + \frac{\beta}{2} \sin \omega \tau) + a_2 (-\omega \sin \omega \tau + \frac{\beta}{2} \cos \omega \tau) \right]. \tag{3.323}$$

From these expressions we get the characteristics

$$a_1 = \frac{e^{-\beta \tau / 2}}{\omega} \left[k(\omega \sin \omega \tau - \frac{\beta}{2} \cos \omega \tau) + r \cos \omega \tau \right] \tag{3.324}$$

$$a_2 = \frac{e^{-\beta \tau / 2}}{\omega} \left[k(\omega \cos \omega \tau + \frac{\beta}{2} \sin \omega \tau) - r \cos \omega \tau \right]. \tag{3.325}$$

Using Eq. (3.323) in the last equation of (3.319), making the integration and the above characteristics, it follows that

$$\hat{\rho}_s(r, k.\tau) = G\big(a_1(\tau), a_2(\tau)\big)e^{-D[(\beta^2 + 4\omega^2)k^2 + 4r^2]/8} , \qquad (3.326)$$

where G is an arbitrary function depending on the characteristics and is determined by the initial conditions. Assume that $\hat{\rho}_s(r, k, 0) = \hat{\rho}_0(r, k)$, then, since $a_1(0) = (r - \beta k/2)/\omega$ and $a_2(0) = k$, one gets

$$\hat{\rho}_0(r, k) = G\big((r - \beta k/2)/\omega, k\big)e^{-D[(\beta^2 + 4\omega^2)k^2 + 4r^2]/8} .$$

Now, defining the parameters $\xi_1 = (r - \beta k/2)/\omega$ and $\xi_2 = k$, we obtain the functionality of G as

$$G(\xi_1, \xi_2) = \hat{\rho}_0\big(\xi_1\omega + \beta\xi_2/2, \xi_2\big)e^{-D[(\beta^2 + 4\omega^2)\xi_2^2 + 4(\xi_1\omega + \beta\xi_2/2)^2]/8} .$$

Thus, the solution of our problem is given by

$$\hat{\rho}_s(r, k, \tau) = \hat{\rho}_0\big(a_1(\tau)\omega + \beta a_2(\tau)/2, a_2(\tau)\big) \times$$
$$e^{-D[(\beta^2 + 4\omega^2 - p_1(\tau))k^2 + (4 - p_2(\tau))r^2 + p_3(\tau)kr]/8} , \qquad (3.327)$$

where one has $p_1(0) = \beta^2 + 4\omega^2$, $p_2(0) = 4$, and $p_3(0) = 0$, and these functions have been defined as

$$p_1(\tau) = \frac{e^{-\beta\tau}(\beta^2 + 4\omega^2)}{4\omega^2}\big[\beta^2 + 4\omega^2 - \beta^2\cos 2\omega\tau + 2\beta\omega\sin 2\omega\tau\big] \quad (3.328)$$

$$p_2(\tau) = \frac{e^{-\beta\tau}}{4\omega^2}\big[4(\beta^2 + 4\omega^2) - 4\beta^2\cos 2\omega\tau - 8\beta\omega\sin 2\omega\tau\big] \qquad (3.329)$$

$$p_3(\tau) = \frac{e^{-\beta\tau}}{4\omega^2}\big[-4\beta(\beta^2 + 4\omega^2) + 4\beta(\beta^2 + 4\omega^2)\cos 2\omega\tau\big] . \qquad (3.330)$$

Once we have gotten the solution in the Fourier space, the solution in the normal space (R) is gotten by using the inverse Fourier transformation,

$$\rho_s(r, R, \tau) = \mathcal{F}^{-1}[\rho_s(r, k, \tau)] = \sqrt{2\pi}\int_{\mathbb{R}} e^{-ikR}\hat{\rho}_s(r, k, \tau)dk.$$

However, one can calculate some dynamical variables without doing this, for example

$$\langle z\rangle(\tau) = tr(\rho_s z) = \int_{\mathbb{R}} z\big[\rho_{1/2}(z, z, \tau) + \rho_{-1/2}(z, z, \tau)\big]\,dz. \qquad (3.331)$$

To see this, note that $\rho_s(z, z, \tau)$ implies from (3.314) that $r = 0$ and $R = z$, that is, it follows that

$$
\begin{aligned}
\langle z \rangle(\tau) &= \int_{\mathbb{R}} R\big[\rho_{1/2}(0, R, \tau) + \rho_{-1/2}(0, R, \tau)\big]\, dR \\
&= \int_{\mathbb{R}} R \mathcal{F}^{-1}\big[\hat{\rho}_{1/2}(0, k, \tau) + \rho_{-1/2}(0, k, \tau)\big]\, dR \\
&= \sqrt{2\pi} \int_{\mathbb{R}} R\, dR \int_{\mathbb{R}} e^{-ikR}\big[\hat{\rho}_{1/2}(0, k, \tau) + \rho_{-1/2}(0, k, \tau)\big]\, dk \\
&= -i\sqrt{2\pi} \int_{\mathbb{R}} dR \int_{\mathbb{R}} e^{-ikR}\left[\frac{\partial \hat{\rho}_{1/2}(0, k, \tau)}{\partial k} + \frac{\partial \rho_{-1/2}(0, k, \tau)}{\partial k}\right]\, dk \\
&= -i\sqrt{2\pi} \left[\frac{\partial \hat{\rho}_{1/2}(0, k, \tau)}{\partial k} + \frac{\partial \rho_{-1/2}(0, k, \tau)}{\partial k}\right]_{k=0}.
\end{aligned}
$$

Therefore, the expectation value of the magnetic flux can be calculated by knowing the reduced density matrix in the Fourier space as

$$
\langle z \rangle(\tau) = -i\sqrt{2\pi} \left[\frac{\partial \hat{\rho}_{1/2}(0, k, \tau)}{\partial k} + \frac{\hat{\partial} \rho_{-1/2}(0, k, \tau)}{\partial k}\right]_{k=0}. \tag{3.332}
$$

Assume that the initial reduced density matrix is given by

$$
\rho_{ss'}^{(0)}(z, z') = \frac{1}{2}\begin{pmatrix} 1 & 1 \\ 1 & 1 \end{pmatrix}\frac{1}{\sqrt{2\pi}} e^{ip_o(z-z') - (z-z_o)^2/2 - (z'-z_o)^2/2}. \tag{3.333}
$$

Then, the diagonal element in the coordinates (r, R) is

$$
\rho_0(r, R) = \frac{1}{2\sqrt{2\pi}} e^{ip_o r - (R+r/2-z-o)^2/2 - (R-r/2-z_o)^2/2}, \tag{3.334}
$$

and its Fourier transformation is given by

$$
\hat{\rho}_0(r, k) = \frac{1}{2\sqrt{2\pi}} e^{ip_o r - r^2/4 + ikz_o - k^2/4}. \tag{3.335}
$$

Writing its dependence in terms of a_1 and a_2, it follows that

$$
\hat{\rho}_0(a_1\omega + \beta a_2/2, a_2) = \frac{1}{2\sqrt{2\pi}} e^{ip_o[a_1\omega + \beta a_2/2] - [a_1\omega + \beta a_2/2]^2/4 + ia_2 z_o - a_2^2/4} \tag{3.336}
$$

since from Eq. (3.332) this expression now must be valuated at $r = 0$, we see from (3.324) and (3.325) that $a_1(r = 0) = \chi_1(\tau)k$ and $a_2(r = 0) = \chi_2(\tau)k$, where

$$
\chi_1(\tau) = \frac{e^{-\beta\tau/2}}{\omega}\left(\omega \sin \omega\tau - \frac{\beta}{2}\cos \omega\tau\right) \tag{3.337}
$$

$$
\chi_2(\tau) = \frac{e^{-\beta\tau/2}}{\omega}\left(\omega \cos \omega\tau + \frac{\beta}{2}\sin \omega\tau\right). \tag{3.338}
$$

In this way, one gets

$$\hat{\rho}_0(a_1\omega + \beta a_2/2, a_2)\big|_{r=0} = \frac{1}{2\sqrt{2\pi}} e^{i\{p_o[\chi_1\omega + \beta\chi_2/2] + z_o\}k - \{[\chi_1\omega + \beta\chi_2/2]^2 + 1\}k^2/4}.$$
(3.339)

Using this expression in Eq. (3.327), and after valuating at $r = 0$ and making its differentiation and valuating at $k = 0$, we get

$$\langle z \rangle(\tau) = z_o + \frac{p_o}{2}\left\{ (\omega\chi_1 + \beta\chi_2/2)_{s=1/2} + (\omega\chi_1 + \beta\chi_2/2)_{s=-1/2} \right\}.$$
(3.340)

Note from (3.322) that ω also depends on s.

Non diagonal elements: Let us write the non diagonal elements of the reduced density matrix as $\hat{\rho}'_s = \hat{f} + i\hat{g}$, and let us separate the real and imaginary parts from the Eq. (3.318). The equations for f and g can be written as

$$L(\hat{f}) = 2\eta s \frac{\partial^2 \hat{g}}{\partial r^2}, \quad L(\hat{g}) = -2\eta s \frac{\partial^2 \hat{f}}{\partial r^2},$$
(3.341)

where L is the linear operator defined as

$$L = \frac{\partial}{\partial \tau} + (\beta r - k)\frac{\partial}{\partial r} + r\frac{\partial}{\partial k} + \left(D\beta r^2 + \frac{sk^2}{2}\right).$$
(3.342)

Eq. (3.341) can be solved by iteration method in any compact set of $\mathbb{R}^3$, related to the variables r, k and τ. Assuming one has this compact set, the equations to be solved are

$$L(\hat{f}_n) = 2\eta s \frac{\partial^2 \hat{g}_{n-1}}{\partial r^2}, \quad L(\hat{g}_n) = -2\eta s \frac{\partial^2 \hat{f}_{n-1}}{\partial r^2}, \quad n \geq 1,$$
(3.343)

and such that $L(\hat{f}_0) = L(\hat{g}_0) = 0$. The equation defined by $L(u) = 0$ is given by

$$\frac{\partial u}{\partial \tau} + (\beta r - k)\frac{\partial u}{\partial r} + r\frac{\partial u}{\partial k} = -\left(D\beta r^2 + \frac{sk^2}{2}\right)u$$
(3.344)

and represents a linear PDEFO defined in $\mathbb{R}^3$, $u = u(r, k, \tau)$. The equations for its characteristics in parametric form are given by

$$\frac{dk}{d\tau} = r, \quad \frac{dr}{d\tau} = \beta r - k, \quad \text{and} \quad \frac{du}{d\tau} = -\left(D\beta r^2 + \frac{sk^2}{2}\right)u.$$
(3.345)

Making the differentiation with respect to τ of the first two equations above, one gets

$$\frac{d^2k}{d\tau^2} + k - \beta\frac{dk}{d\tau} = 0$$
(3.346)

which has the solution

$$k(\tau) = e^{\beta\tau/2}\big(b_1 \sin\omega_o\tau + b_2 \cos\omega_o\tau\big), \qquad (3.347)$$

where ω_o has been defined as

$$\omega_o = \sqrt{1 - \beta^2/4}. \qquad (3.348)$$

In this way, the solution for the variable r is

$$r(\tau) = e^{\beta\tau/2}\Big[(\frac{\beta}{2}\sin\omega_o\tau + \omega_o\cos\omega_o\tau)b_1 + (\frac{\beta}{2}\cos\omega_o\tau - \omega_o\sin\omega_o\tau)b_2\Big].$$
$$(3.349)$$

From these expressions we get the characteristics

$$b_1 = \frac{e^{-\beta\tau/2}}{\omega_o}\Big[k(\omega_o\sin\omega_o\tau - \frac{\beta}{2}\cos\omega_o\tau) + r\cos\omega_o\tau\Big] \qquad (3.350)$$

$$b_2 = \frac{e^{-\beta\tau/2}}{\omega_o}\Big[k(\omega_o\cos\omega_o\tau + \frac{\beta}{2}\sin\omega_o\tau) - r\cos\omega_o\tau\Big] \qquad (3.351)$$

making use of these expressions, the integration of the last term in 3.345 can be done, bringing about the solution

$$u(r,k,\tau) = G(b_1, b_2)e^{-[\alpha_1 k^2 + \alpha_2 r^2 - 2\alpha_3 kr]}, \qquad (3.352)$$

where G is an arbitrary function depending on the characteristics, and α_i for $i = 1, 2, 3$ has been defined as

$$\alpha_1 = \frac{D}{8}(\beta^2 + 4\omega_o^2) + \frac{s(5\beta^2 + 4\omega_o^2)}{4\beta(\beta^2 + 4\omega_o^2)} \qquad (3.353)$$

$$\alpha_2 = \frac{D}{2} + \frac{s}{\beta^2 + 4\omega_o^2} \qquad (3.354)$$

$$\alpha_3 = \frac{s}{\beta^2 + 4\omega_o^2}. \qquad (3.355)$$

The functionality of G is determined by the initial conditions, $u(r,k,0) = u_o(r,k)$, and proceeding in the same way as we did for the diagonal case, one gets

$$G(\xi_1, \xi_2) = u_o(\omega_o\xi_1 + \beta\xi_2/2, \xi_2)e^{\alpha_1\xi_2^2 + \alpha_2(\omega_o\xi_1 + \beta\xi_2/2)^2 - 2\alpha_3\xi_2(\omega_o\xi_1 + \beta\xi_2/2)},$$

where ξ_1 and ξ_2 are any variables. Therefore, the solution of Eq. (3.344) can be written (after some arrangements) as

$$u(r,k,\tau) = u_o\big(\omega_o b_1(\tau) + \beta b_2(\tau)/2, b_2(\tau)\big) \times$$
$$e^{\{(\alpha_1 - q_1)k^2 + (\alpha_2 - q_2)r^2 - 2(\alpha_3 + q_3)kr\}}, \qquad (3.356)$$

where the functions q_i for $i = 1, 2, 3$ have been defined as

$$
q_1(\tau) = \frac{e^{-\beta\tau}}{16\omega_o}\Big\{\big(4\beta^2\alpha_1 + \beta^4\alpha_2 - 4\beta^3\alpha_3 + 8\beta^2\omega_o^2\alpha_2 - 16\beta\omega_o^3\alpha_3
$$
$$
+16\omega^2\alpha_2\big)\sin^2\omega_o\tau + 16\omega_o^2\alpha_1\cos^2\omega_o\tau + \big(8\beta\omega_o\alpha_1 - 4\beta^2\omega_o\alpha_3
$$
$$
-16\omega_o^3\alpha_3\big)\sin 2\omega_o\tau\Big\} \tag{3.357}
$$

$$
q_2(\tau) = \frac{e^{-\beta\tau}}{16\omega_o^2}\Big\{16\omega_o^2\alpha_2\cos^2\omega_o\tau + \big(16\alpha_1 + 4\beta^2\alpha_2 - 16\beta\alpha_3\big)\sin^2\omega_o\tau
$$
$$
+\big(16\omega_o\alpha_3 - 8\beta\omega_o\alpha_2\big)\sin 2\omega_o\tau\Big\} \tag{3.358}
$$

$$
q_3(\tau) = \frac{e^{-\beta\tau}}{16\omega_o^2}\Big\{-16\omega_o^2\alpha_3\cos^2\omega_o\tau + \big(-2\beta^2\alpha_2 + 8\beta^2\alpha_3 + 16\omega_o^2\alpha_3
$$
$$
-8\beta(\alpha_1 + \alpha_2\omega_o^2)\big)\sin^2\omega_o\tau - 2\omega_o\big(-\beta^2\alpha_2 + 4(\alpha_1 - \alpha_2\omega_o^2)
$$
$$
\times \sin 2\omega_o\tau\Big\}. \tag{3.359}
$$

These functions have the initial values

$$
q_1(0) = \alpha_1, \quad q_2(0) = \alpha_2, \quad \text{and} \quad q_3(0) = -\alpha_3. \tag{3.360}
$$

Having solved the problem $L(u) = 0$, we can now proceed to solve Eq. (3.343) for any $n \geq 1$. Let us solve first the case for $n = 1$, and the generalization will be given straightforwardly. The equation defines by (3.343) is written as

$$
\frac{\partial \hat{f}_1}{\partial \tau} + (\beta r - k)\frac{\partial \hat{f}_1}{\partial r} + r\frac{\partial \hat{f}_1}{\partial k} = -\Big(D\beta r^2 + \frac{sk^2}{2}\Big)\hat{f}_1 + 2\eta s\frac{\partial u}{\partial r^2}. \tag{3.361}
$$

This expression represents a linear PDEFO defined in $\mathbb{R}^3$, where the equations for its characteristics are given by Eq. (3.345) but the last one which is changed by

$$
\frac{d\hat{f}_1}{d\tau} = -\Big(D\beta r^2 + \frac{sk^2}{2}\Big)\hat{f}_1 + 2\eta s\frac{\partial^2 u}{\partial r^2}. \tag{3.362}
$$

The solution for r and k is the same as Eq. (3.347) and Eq. (3.349), with the same angular frequency (3.348) and the same characteristics (3.350) and (3.351). The solution of Eq. (3.362) is readily given by

$$\hat{f}_1(r,k,\tau) = \widetilde{G}(b_1,b_2)e^{-[\alpha_1 k^2 + \alpha_2 r^2 - 2\alpha_3 kr]}$$
$$+2\eta s e^{-[\alpha_1 k^2 + \alpha_2 r^2 - 2\alpha_3 kr]} \int_0^\tau \left(\frac{\partial^2 u}{\partial r^2}\right) e^{-[\alpha_1 k^2(\sigma) + \alpha_2 r^2(\sigma) - 2\alpha_3 k(\sigma) r(\sigma)]} \, d\sigma \, .$$

$$(3.363)$$

The arbitrary function $\widetilde{G}$ is determined as before. Thus, the solution of Eq. (3.361) is gotten as

$$\hat{f}_1(r,k,\tau) = u_o\big(\omega_o b_1 + \beta b_2/2, b_2\big)e^{-[(\alpha_1 - q_1)k^2 + (\alpha_2 - q_2)r^2 - 2(\alpha_3 + q_3)kr]}$$
$$+2\eta s e^{-[\alpha_1 k^2 + \alpha_2 r^2 - 2\alpha_3 kr]} \int_0^\tau \left(\frac{\partial^2 u}{\partial r^2}\right) e^{-[\alpha_1 k^2(\sigma) + \alpha_2 r^2(\sigma) - 2\alpha_3 k(\sigma) r(\sigma)]} \, d\sigma$$

$$(3.364)$$

where the functions $q_i = q_i(\tau)$ for $i = 1, 2, 3$ are given as above. A similar solution (changing η to $-\eta$) is obtained for the imaginary part $\hat{g}_1$ of Eq. (3.341). In addition, It is straightforward to see the general solution of Eq. (3.343) which can be written as

$$\hat{f}_n(r,k,\tau) = u_o\big(\omega_o b_1 + \beta b_2/2, b_2\big)e^{-[(\alpha_1 - q_1)k^2 + (\alpha_2 - q_2)r^2 - 2(\alpha_3 + q_3)kr]}$$
$$+2\eta s e^{-[\alpha_1 k^2 + \alpha_2 r^2 - 2\alpha_3 kr]} \int_0^\tau \left(\frac{\partial^2 \hat{g}_{n-1}}{\partial r^2}\right) e^{-[\alpha_1 k^2(\sigma) + \alpha_2 r^2(\sigma) - 2\alpha_3 k(\sigma) r(\sigma)]} \, d\sigma$$

$$(3.365)$$

and

$$\hat{g}_n(r,k,\tau) = \tilde{u}_o\big(\omega_o b_1 + \beta b_2/2, b_2\big)e^{-[(\alpha_1 - q_1)k^2 + (\alpha_2 - q_2)r^2 - 2(\alpha_3 + q_3)kr]}$$
$$-2\eta s e^{-[\alpha_1 k^2 + \alpha_2 r^2 - 2\alpha_3 kr]} \int_0^\tau \left(\frac{\partial^2 \hat{f}_{n-1}}{\partial r^2}\right) e^{-[\alpha_1 k^2(\sigma) + \alpha_2 r^2(\sigma) - 2\alpha_3 k(\sigma) r(\sigma)]} d\sigma \, .$$

$$(3.366)$$

These solutions are given in the Fourier space, and one still need to make the inverse transformation to get the solution in our variables (r, R, τ). Since the solution for any "n" depends on the first solution f_1 and g_1, the decoherence behavior ($\rho'_s \to 0$ for $s = 1/2, -1/2$) depends mainly on the decoherence of these first elements ($f_1 \to 0$ and $g_1 \to 0$).

3.14 Problems

3.1 Write down the PDEFO associated to the constant of motion, K, of the dynamical system $dx/dt = v$ and $dv/dt = (-g + \alpha v^2/m)$, where α is a non negative constant, and m represents the mass of a particle.

i) Show that the characteristic can be written as

$$C = -\frac{mg}{2\alpha} \ln\left(1 - \frac{\alpha v^2}{mg}\right) + gx.$$

ii) Show that the solution of the equation for the constant of motion is given as $K = G(C)$, where G is a general function.

iii) Show that by choosing $G(C) = mC$ and $G(C) = -(mg/2\alpha)e^{-2\alpha C/mg} - m^2g/2\alpha$, one gets the following two equivalents constant of motions

$$K_1(x,v) = -\frac{m^2g}{2\alpha} \ln\left(1 - \frac{\alpha v^2}{mg}\right) + mgx$$

and

$$K_2(x,v) = \frac{m^2}{2\alpha}\left(-g + \frac{\alpha v^2}{m}\right)e^{-2\alpha x/m} + \frac{m^2g}{2\alpha}.$$

iv) Show that the following limit is gotten $\lim_{\alpha\to 0} K_i = mv^2/2 + mgx$ for $i = 1, 2$.

v) Show that we can get two Lagrangian for the system given by

$$L_1(x,v) = m\sqrt{\frac{mg}{\alpha}}\, v\, \text{arctanh}\left(v\sqrt{\frac{\alpha}{mg}}\right) + \frac{m^2g}{2\alpha} \ln\left(1 - \frac{\alpha v^2}{mg}\right) - mgx$$

and

$$L_2(x,v) = \frac{m^2}{2\alpha}\left(g + \frac{\alpha v^2}{m}\right)e^{-2\alpha x/m} - \frac{m^2g}{2\alpha}.$$

vi) Show that $\lim_{\alpha\to 0} L_i = mv^2/2 - mgx$ for $i = 1, 2$.

vii) Show that, given these two Lagrangian, one gets the following two Hamiltonian for the system

$$H_{(x,p)} = -\frac{m^2g}{2\alpha} \ln\left[1 - \tanh^2\left(\frac{p}{m}\sqrt{\frac{\alpha}{mg}}\right)\right] + mgx$$

and

$$H_2(x,p) = \frac{m^2}{2\alpha}\left(-g + \frac{\alpha p^2}{m^3}e^{4\alpha x/m}\right)e^{-2\alpha x/m} + \frac{m^2g}{2\alpha}.$$

viii) Show that $\lim_{\alpha\to 0} H_i = p^2/2m + mgx$ for $i = 1, 2$.

3.2 Find the solution of the PDEFO found in exercise (12) of chapter 1, assuming that the equation

$$\frac{dx}{v} = \frac{dv}{F(x,v)}$$

brings about the characteristic curve $C = C(x,v)$, and show that the solution for $L(x,v)$ $(J = L_{vv})$ can be written as

$$L(x,v) = \int^v dv' \int^{v'} dv''\, A\big(C(x,v'')\big) e^{-\int^{v''} F_{\bar{v}}\, d\bar{v}/F(x,\bar{v})} + B_1 v + B_2\ ,$$

where B_1 and B_2 are constants, and A is an arbitrary function of the characteristic C.

3.3 The equation of motion of a one-degree of freedom relativistic particle of mass m in a dissipative medium characterized by a quadratic dissipation force with dissipation constant α, $-\alpha(dx/dt)^2$ and $v \geq 0$, can be written in Newton form as

$$m\frac{d^2x}{dt^2} = -\alpha v^2 (1 - v^2/c^2)^{3/2}\ , \qquad c \text{ is the speed of light}\ .$$

i) Find the constant of motion associated to this system by solving its associated PDEFO.
ii) Determine this constant of motion such that $\lim_{c\to\infty} K = mv^2/2$.
iii) If it is possible, find the Lagrangian of the system by using Eq. (3.17a). If it is not possible, determine the main difficulty.
iv) If it is possible, find the Hamiltonian of the system by using Eq. (3.29). If it is not possible, determine the main difficulty.
v) Since $v/c \leq 1$, write down the constant of motion in Taylor series expansion around $v/c = 0$, and using this expression, repeat the steps (iii) and (iv).

3.4 Show that the constant of motion associated to the dynamical system

$$\frac{dx}{dt} = v\ , \qquad \frac{dv}{dt} = -\omega^2 x - \frac{\alpha}{m}v$$

can be given by the following expression

$$K(x,v) = \frac{m}{2}\left(v^2 + 2\omega_\alpha x v + \omega^2 x^2\right) e^{-2\omega_\alpha G_\alpha(v/x,\omega)}\ ,$$

where α is the dissipative constant of the system, ω_α is defined as $\omega_\alpha = \alpha/2m$, and G_α is defined as

$$G_\alpha = \begin{cases} \dfrac{1}{2\sqrt{\omega_\alpha^2-\omega^2}} \ln\left(\dfrac{\omega_\alpha+v/x-\sqrt{\omega_\alpha^2-\omega^2}}{\omega_\alpha+v/x+\sqrt{\omega_\alpha^2-\omega^2}}\right), & \text{if } \omega^2 < \omega_\alpha^2 \\[3mm] \dfrac{1}{\omega^2+v/x}, & \text{if } \omega^2 = \omega_\alpha^2 \\[3mm] \dfrac{1}{\sqrt{\omega^2-\omega_\alpha^2}} \arctan\left(\dfrac{\omega_\alpha+v/x}{\sqrt{\omega^2-\omega_\alpha^2}}\right), & \text{if } \omega^2 > \omega_\alpha^2 \end{cases}$$

corresponding to the weak, critical, and strong dissipation cases. Show the difficulty to find the Lagrangian and the Hamiltonian of the system, even at first order in the parameter ω_α (very weak dissipation).

3.5 Find the eigenvalues of the operator L_+ (Eq. (3.41a)), and look for a polynomial solution.

3.15   *References*

3.1   G. López, Constant of Motion, Hamiltonian, and Lagrangian for Autonomous Systems in Hyperbolic Flat Spaces, SSCL-Preprint-150, August 1993.

3.2   A.R. Edmonds, *Angular Moment in Quantum Mechanics*, Princeton University Press, 1957. Chap. 2.

3.3   B.L. Vander Waerden, *The Group Method in Quantum Mechanics*, Springer-Verlag, 1938. Chap. 3.

3.4   G. López, *Analytical Approximation to the Turn-to-Turn Quench Propagation*, SSCL-309, September 1990.

3.5   K. Huang, *Statistical Mechanics*, John Wiley & Sons, 1987. Chap.6-7.

3.6   P. Srivastava, *TCP (truncated compound Poisson) Process for Multiplicity Distribution in High Energy Collisions*, International Center for Theoretical Physics, Ic/88/127.

3.7   E.D. Courant and H.S. Snyder, Ann. Phys., **3** (1958) 1.

3.8   M. Sands, *The Physics of Electron Storage Rings, and Introduction*, SLAC-121, Nov. 1970.

3.9   G. López and S. Chen, *Tune Shift Effect Due to Multipole Longitudinal Periodic Structure in the Superconducting Dipole Magnets*, SSCL-550, October 1991.

3.10   H. Goldstein, *Classical Mechanics*, Addison-Wesley, Reading MA, 1950.

3.11   A.W. Chao, *Physics of Collective Beam Instabilities in High Energy Accelerators*, John Wiley and Sons, Inc., 1993.

3.12   P.J. Bryan and K. Johnsen, *The Principles of Circular Accelerators and Storage Rings*, Cambridge University Press, 1993.

3.13   A.W. Chao, *Coherent Instabilities of a Relativistic Bunched Beam*, SLAC-PUB-2946, June 1982.

3.14   G. López, Ann. Phys., **251**, (1996) 1.

3.14   N.M. Atakishiyev, T.H. Seligman, K.B. Wolf, *Proceedings of the IV Wigner Symposium*, World Scientific, 1995. Page 250.

3.15   A.N. Kolmogorov, Dokl. Akad. Nauk. SSSR, **98**, (1954) 527.

3.16   V.I. Arnold, Dokl. Akad. Nauk. SSSR, **156**, (1964) 9.

3.17   L. Landau, *On the vibration of the electronic plasma.* J. Phys. USSR **10**,25, (1946). (JETP **16**, 574, (1946)).

3.18   V. Maslov and M. Fedoryuk, *The linear theory of Landau damping*, Mat. Sb. 127, 445 (1985).

3.19   J.R. Reitz, *Foundation of Electromagnetic Theory*, Addison Wesley, (1969), chapter 14.

3.20   J. Scheffield, *Plasma Scattering of Electromagnetic Radiation*, Academic Press (1975), appendix 2.

3.21   S. Chandrasekhar,*Plasma Physics*, The University of Chicago Press (1960), chapter VII.

3.22   J.D. Jackson,*Classical Electrodynamics* (1999), chapter 14.

3.23   C. Cercignani, *The Boltzmann Equation and Its Applications*, Springer-Verlag (1988).

3.24   A.O. Benz, *Plasma Astrophysics, Kinetic Processes in Solar and Stellar Coronae*, Kluwer Academic Publishers, (2002), chapter 5.

3.25   A. Messiah, *Quantum Mechanics*, John Wiley & Sons, (1976), vol. I, II.

3.26   P.A.M. Dirac, *The Principles of Quantum Mechanics*, Oxford University Press, (1976).

3.27   G. Lindblad, *On the generators of quantum dynamical semigroups*, Commun. Math. Phys. **48**, (1976) 119.

3.28   A.O. Caldeira and A.J. Legget, *Path integral approach to quantum Brownian motion*, Physica A, **121** (1983) 587.

3.29   G. Johanson, L. Tornberg, and C.M. Wilson, Phys. Rev. B., **74**, (2006) 100504(R).

3.30   H.P. Breuer and F. Petruccione, *The Theory of Open Quantum Systems*, Oxford University Press, (2009).

3.31 Y. Makhlin, G. Schön, and A. Shnirman, Rev. Mod. Phys., **73**, (2001) 357.

3.32 G. López and P. López, internal report, University of Guadalajara, México,RI-2009-COC1, (2009).

Chapter 4

Nonlinear Partial Differential Equations of First Order

In this chapter we will developed the theory of the nonlinear PDEFO defined on $\mathbb{R}^2$ and on $\mathbb{R}^n$.

4.1 Non-Linear PDEFO for Functions Defined in $\mathbb{R}^2$

The PDEFO for functions defined in the plane (x, y) have the general form

$$F(x, y, z, q, p) = 0, \qquad (4.1)$$

where $z = z(x, y)$, and p and q are defined as

$$p = \frac{\partial z}{\partial x}, \quad q = \frac{\partial z}{\partial y}. \qquad (4.2)$$

Let $z = f(x, y)$ be the integral surface of Eq. (4.1), then at any point $P = (x, y, z)$ on this surface, Eq. (4.1) gives us a nonlinear relation between p and q

$$\phi(p, q) = 0 \qquad (4.3)$$

in such a way that there can exist more than one possible solution for p, with q given (or the other way around), then the normal vector,

$$\hat{n} = \frac{1}{(p^2 + q^2 + 1)^{1/2}} (p, q, -1), \qquad (4.4)$$

to the surface at this point is not unique defined (see Fig. 4.1). In this way we obtain a one parameter family of normal-vectors to the surfaces at the point P. Its envelope will be called the Monges's cone of the normal directions. As a consequence, a one parameter family of tangent planes passing through the point $P(x, y, z)$ is formed, $Z - z = p(X - x) + q(Y - y)$, and this envelope is called Monge's cone of the tangent planes.

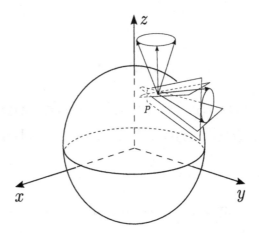

Fig. 4.1   Family of normal vectors generated by (4.4), the envelope is generated by the non linear relation between p and q (4.2).

Let establish this point with particular examples. In the quasi-linear case, see Eq. (2.15), we have

$$F(x, y, z, q, p) = a(x, y, z)p + b(x, y, z)q - c(x, y, z) = 0,$$

evaluating the coefficients a, b and c at a given point P, they become real numbers and the above differential equation becomes a linear algebraic equation between p and q. Then, knowing q at the point $P(x, y, z)$, there exist one and only one value for p, and this defines a unique normal direction. However, consider the non-linear partial differential equation

$$F(x, y, z, p, q) = a_n p^n + a_{n-1} p^{n-1} + ... + a_1 p + bq - c = 0$$

where the coefficients a_i $i = 1, ..., n$, b and c are arbitrary functions at x, y and z. Evaluating the above equation at the point $P(x, y, z)$, becomes an algebraic equation of order n in p and knowing q there exists n-complex possible values for p, that is, the normal of the surface at the point $P(x, y, z)$ is not uniquely defined.

Assume that we have the integral surface of Eq. (4.1) given by

$$\Phi(x, y, z, a, b) = 0, \tag{4.5}$$

where a and b are constants. The relation (4.5) defines a two parameters family of surfaces. The envelope of this family will be also an integral surface because the normal vector to the envelope surface coincides with the normal vector to one of the integral surface of the family. The determination of this envelope is given by

$$\Phi(x, y, z, a, b) = 0, \quad \frac{\partial \Phi}{\partial a} = 0, \quad \frac{\partial \Phi}{\partial b} = 0. \tag{4.6}$$

We can obtain a one parameter of integral surface if we make $b = b(a)$ and in this case the envelope will be determined by

$$\Phi(x, y, z, a, b(a)) = 0, \quad \frac{\partial}{\partial a}\Phi(x, y, z, a, b(a)) = 0 \tag{4.7a}$$

or

$$\Phi(x, y, z, a, b(a)) = 0, \quad \frac{\partial \Phi}{\partial a} + \frac{\partial \Phi}{\partial b}\frac{\partial b}{\partial a} = 0. \tag{4.7b}$$

Because these last relation contain an arbitrary parameter, this allows us to obtain an integral surface which satisfies the Cauchy initial conditions.

Example 4.1. Find the envelope to the family of surfaces given by the equation

$$(x - a)^2 + y^2 - z^2 = b^2.$$

The family of surfaces (4.5) is

$$\Phi(x, y, z, a, b) = (x - a)^2 + y^2 - z^2 - b^2 = 0.$$

The set of equations (4.6) are then

$$(x - a)^2 + y^2 - z^2 - b^2 = 0, \quad -2(x - a) = 0, \quad -2b = 0,$$

which implies that the straight lines defined by

$$z = \pm y$$

are the envelopes.

Let us see first some particular cases where Eq. (4.1) becomes simple to solve.

Case (a): Suppose that the PDEFO is of the form

$$F(p, q) = 0 . \tag{4.8a}$$

This expression implies that

$$p = \phi(q). \tag{4.8b}$$

Making $q = a$, where a in an arbitrary constant, we get from Eq. (4.8a)

$$p = \phi(a),$$

then, it follows that

$$dz = pdx + qdy = \phi(a)dx + ady,$$

which easily can be integrated, obtaining

$$z = \phi(a)x + ay + b. \tag{4.9}$$

Example 4.2. Find the integral surface of the following equation

$$\left(\frac{\partial z}{\partial x}\right)^2 - \left(\frac{\partial z}{\partial y}\right)^2 = 0.$$

Making $q = \partial z/\partial y = a$, we have that $p = \partial z/\partial x = \pm a$, then the solution is

$$z = \pm ax + ay + b.$$

Case (b): Suppose that the PDEFO can be written in the form

$$f_1(x, p) = f_2(y, q). \tag{4.10}$$

Making

$$f_1(x, p) = f_2(y, q) = a, \tag{4.11}$$

where "a" is an arbitrary constant, and if we can solve Eq. (4.11) with respect to p and q, that is

$$p = \phi_1(x, a)$$

and

$$q = \phi_2(y, a),$$

then we have

$$dz = pdx + qdy = \phi_1(x, a)dx + \phi_2(y, a)dy$$

which can be integrated, giving the solution

$$z = \int \phi_1(x, a)dx + \int \phi_2(y, a)dy + b. \tag{4.12}$$

Example 4.3. Find the integral surfaces of the equation

$$y^2 \left(\frac{\partial z}{\partial x}\right)^3 - x^3 \left(\frac{\partial z}{\partial y}\right)^2 = 0.$$

This equation can be written in the form (4.10), then we can make

$$x^{-3} \left(\frac{\partial z}{\partial x}\right)^3 = y^{-2} \left(\frac{\partial z}{\partial x}\right)^2 = a,$$

and from these relations, we have

$$p = a^{1/3}x, \quad q = \pm a^{1/2}y.$$

Then the solution of Eq. (4.12) is

$$z = \frac{1}{2}a^{1/3}x^2 \pm \frac{a^{1/2}}{2}y^2 + b.$$

Case (c): Suppose that the PDEFO is of the form

$$F(z, p, q) = 0. \tag{4.13}$$

Making $z = z(u)$ where $u = ax + y$, Eq. (4.13) is transformed to an ordinary differential equation,

$$F\left(z, a\frac{dz}{du}, \frac{dz}{du}\right) = 0, \tag{4.14}$$

which can be integrated, obtain a solution of the form

$$z = \Phi(ax + y, a, b).$$

Example 4.4. Find the integral surfaces of the equation

$$z = \left(\frac{\partial z}{\partial x}\right)\left(\frac{\partial z}{\partial y}\right).$$

The new variable $u = ax + y$ transform this equation to

$$z = a\left(\frac{dz}{du}\right)^2.$$

Integrating this equation brings about the solution

$$z = \frac{1}{4}\left(a^{1/2}x + a^{-1/2}y + b\right)^2. \tag{4.15}$$

Case (d): If the PDEFO can be written in the form

$$z = px + qy + f(q, p), \tag{4.16}$$

called Clairaut's equation, we can make $p = a$, $q = b$, having the differential

$$dz = adx + bdy,$$

which has the solution

$$z = ax + by + f(a, b). \tag{4.17}$$

Example 4.5. Find the integral surfaces of the equation

$$z = x\left(\frac{\partial z}{\partial x}\right) + y\left(\frac{\partial z}{\partial y}\right) + \left(\frac{\partial z}{\partial x}\right)^2\left(\frac{\partial z}{\partial y}\right).$$

The equation is of the form (4.16), then the solution is

$$z = ax + by + a^2b.$$

Let us see now the Lagrange-Charpit method for the solution of the PDEFO given by Eq. (4.1). Suppose we find a function $U(x, y, z, p, q)$ such that, from the equations

$$F(x, y, z, p, q) = 0 \qquad (4.18a)$$

and

$$U(x, y, z, p, q) = a, \qquad (4.18b)$$

where "a" ia an arbitrary constant and $\partial(F, U)/\partial(p, q) \neq 0$, we can have

$$p = p(x, y, z, a) \qquad (4.19a)$$

and

$$q = q(x, y, z, a). \qquad (4.19b)$$

Therefore we need only to integrate

$$dz = p(x, y, z, a)dx + q(x, y, z, a)dy \qquad (4.20)$$

to obtain the two parameters solution

$$\Phi(x, y, z, a, b) = 0. \qquad (4.21)$$

The function U is determined from the integrability condition of the Pfaffian equation (4.20), which defines the vector field $\mathbf{E} = (p, q, -1)$, and this integrability ($\mathbf{E} \cdot \nabla \times \mathbf{E} = 0$), as it was already seen in Eq. (1.54), is given by

$$p\frac{\partial q}{\partial z} - q\frac{\partial p}{\partial z} - \frac{\partial p}{\partial y} + \frac{\partial q}{\partial x} = 0. \qquad (4.22)$$

Thus, if we can find a function U satisfying Eq. (4.22), assuming we also have relations (4.19), the integration of Eq. (4.20) can be made obtaining the solution (4.21). As it was already pointed out, the existence of relations (4.19) can be assured by asking that the following relation be satisfied

$$\frac{\partial(F, U)}{\partial(p, q)} \neq 0 \qquad (4.23)$$

and by the implicit function theorem (see reference T.M. Apostol). Let us calculate the partial derivatives $\partial q/\partial z$, $\partial p/\partial z$, $\partial p/\partial y$ and $\partial q/\partial x$ using relations (4.18). Doing the total derivation with respect x of Eq. (4.18), we have

$$\frac{\partial F}{\partial x} + \frac{\partial F}{\partial p}\frac{\partial p}{\partial x} + \frac{\partial F}{\partial q}\frac{\partial q}{\partial x} = 0$$

and

$$\frac{\partial U}{\partial x} + \frac{\partial U}{\partial p}\frac{\partial p}{\partial x} + \frac{\partial U}{\partial q}\frac{\partial q}{\partial x} = 0$$

which can be written as

$$\begin{pmatrix} F_p & F_q \\ U_p & U_q \end{pmatrix}\begin{pmatrix} \partial p/\partial x \\ \partial q/\partial x \end{pmatrix} = -\begin{pmatrix} F_x \\ U_x \end{pmatrix}.$$

Due to conditions (4.23), the matrix can be inverted. Solving this equation, if follows

$$\begin{pmatrix} \partial p/\partial x \\ \partial q/\partial x \end{pmatrix} = \frac{-1}{\partial(F,U)/\partial(p,q)}\begin{pmatrix} U_q F_x & -F_q U_x \\ -U_p F_x & F_p U_x \end{pmatrix} = \begin{pmatrix} -\frac{\partial(U,F)}{\partial(q,x)}\Big/\frac{\partial(U,F)}{\partial(q,p)} \\ \frac{\partial(U,F)}{\partial(p,x)}\Big/\frac{\partial(U,F)}{\partial(q,p)} \end{pmatrix}.$$

(4.24)

In a similar way, after making the total derivation of Eq. (4.18) with respect to y and z (the index "x" changes to "y" and "z"), we obtain

$$\frac{\partial p}{\partial y} = -\frac{\partial(F,U)}{\partial(y,q)}\Big/\frac{\partial(F,U)}{\partial(p,q)},$$

(4.25)

$$\frac{\partial p}{\partial z} = -\frac{\partial(F,U)}{\partial(z,q)}\Big/\frac{\partial(F,U)}{\partial(p,q)},$$

(4.26a)

and

$$\frac{\partial q}{\partial z} = -\frac{\partial(F,U)}{\partial(p,z)}\Big/\frac{\partial(F,U)}{\partial(p,q)}.$$

(4.26b)

Substituting Eq. (4.24), Eq. (4.25) and Eq. (4.26) in Eq. (4.22) we get

$$p\left[-\frac{\partial(F,U)}{\partial(p,z)}\right] - q\left[-\frac{\partial(F,U)}{\partial(z,q)}\right] + \frac{\partial(F,U)}{\partial(y,q)} - \frac{\partial(F,U)}{\partial(p,x)} = 0$$

or

$$F_p\frac{\partial U}{\partial x} + F_q\frac{\partial U}{\partial y} + (pF_p + qF_q)\frac{\partial U}{\partial z}$$
$$- (F_x + pF_z)\frac{\partial U}{\partial p} - (F_y + qF_z)\frac{\partial U}{\partial q} = 0, \quad (4.27)$$

which is a homogeneous linear PDEFO. Thus, From the equations for this characteristic curves

$$\frac{dx}{F_p} = \frac{dy}{F_q} = \frac{dz}{pF_p + qF_q} = \frac{dp}{-(F_x + pF_z)} = \frac{dq}{-(F_y + qF_z)},$$

(4.28)

we can derive at least one characteristic surface

$$U_1(x,y,z,p,q) = a,$$

(4.29)

which we can use as our Eq. (4.18b), if it satisfies

$$\frac{\partial(F,U)}{\partial(p,q)} \neq 0. \tag{4.30}$$

Therefore, once we have found the pair of relations (4.18), we can obtain Eq. (4.19) and integrating Eq. (4.20), we get the solution Eq. (4.21).

Example 4.6. Find the integral surfaces of the equation

$$yz\left(\frac{\partial z}{\partial x}\right)^2 - \left(\frac{\partial z}{\partial y}\right) = 0.$$

The function F is given by

$$F(x,y,z,p,q) = yzp^2 - q = 0, \tag{4.31}$$

and the partial derivatives are

$$F_p = 2yzp, \quad F_q = -1, \quad F_z = yp^2, \quad F_x = 0, \quad \text{and} \quad F_y = zp^2.$$

The equation for the characteristic (4.28) are

$$\frac{dx}{2pyz} = -dy = \frac{dz}{2p^2yz - q} = \frac{dp}{-yp^3} = \frac{dq}{-(zp^2 + yp^2q)}.$$

From Eq. (4.31), one gets $q = yzp^2$ which can be used in the third term above, and using the fourth term, we get

$$\frac{dz}{yp^2z} = \frac{dp}{-p^3y}$$

which has the solution

$$U_1 = pz = a. \tag{4.32}$$

With Eq. (4.31) and Eq. (4.32), we obtain

$$p = a/z$$

and

$$q = ya^2/z.$$

In this way, the Pfaffian equation (4.20) is

$$dz = \frac{a}{z}dx + \frac{ya^2}{z}dy$$

and can be integrated, giving the solution

$$z^2 = 2ax + a^2y^2 + b.$$

Now, the problem of finding the integral surfaces passing through a given initial curve determined by

$$x = x(t), \quad y = y(t), \quad \text{and} \quad z = z(t), \qquad (4.33)$$

can be solved finding first the function

$$b = b(a) \qquad (4.34)$$

in such a way that the envelope of the one-parameter family of surfaces

$$\Phi(x, y, z, a, b(a)) = 0 \qquad (4.35)$$

determined by Eq. (4.35) and

$$\frac{\partial \Phi}{\partial a} + \frac{\partial \Phi}{\partial b} \frac{\partial b}{\partial a} = 0 \qquad (4.36)$$

passes through the curve define by Eq. (4.33).

The functionality $b = b(a)$ is easier determined by Eq. (4.35) and

$$\frac{\partial \Phi}{\partial x} \frac{\partial x}{\partial t} + \frac{\partial \Phi}{\partial y} \frac{\partial y}{\partial t} + \frac{\partial \Phi}{\partial z} \frac{\partial z}{\partial t} = 0. \qquad (4.37)$$

This means that the tangent vector of the initial curve must be orthogonal to the normal vector of the surface $\Phi = 0$. Once $b(a)$ is determined, this solution is obtained by finding the enveloped of the one parameter family of surfaces already deduced.

Example 4.7. Find the integral surface of the Example 4.4 which passes through the curve

$$x(t) = 1, \quad y(t) = t \quad \text{and} \quad z(t) = t.$$

The two parameter family of integral surface is given by $4z = (a^{1/2}x + a^{-1/2}y + b)^2$, which defines $\Phi(x, y, z, a, b) = (a^{1/2}x + a^{-1/2}y + b)^2 - 4z$. Then, the Eq. (4.35) and Eq. (4.37) are given on the initial curve as

$$\left(a^{1/2} + a^{-1/2}t + b(a) \right)^2 - 4t = 0$$

and

$$2 \left(a^{1/2} + a^{-1/2}t + b(a) \right) a^{-1/2} - 4 = 0.$$

From where we obtain

$$a = t, \quad \text{and} \quad b(a) = a^{1/2} - a^{-1/2}a = 0. \qquad (4.38)$$

Thus, we have a one-parameter family of integral surfaces given by

$$\Phi = \left(a^{1/2}x + a^{-1/2}y\right)^2 - 4z = 0. \tag{4.39}$$

The envelope of these surfaces is determined by Eq. (4.39) and $\partial\Phi/\partial a = 0$, from this one, we obtain

$$\left(a^{1/2}x + a^{-1/2}y\right)\left[\frac{x}{a^{1/2}} - \frac{y}{a^{3/2}}\right] = 0. \tag{4.40a}$$

From this equation, the value of "a" is determined as

$$a = y/x,$$

and substituting this result in Eq. (4.39), we obtain finally the solution

$$z = yx. \tag{4.40b}$$

If the system (4.28) can be integrated, then the solve the Cauchy problems the following method called "the Characteristics", "Cauchy", or sometimes "First Jacobi" method is very useful. Taking s as the parameter for the family curves Γ_s

$$x_0 = x_0(s), \quad y_0 = y_0(s) \quad \text{and} \quad z_0 = z_0(s), \tag{4.41}$$

the integral surface $z = z(x,y)$ passing through the curve can be imagined as being formed by the set of point the certain monoparametric parameter family of curves Γ, called characteristic (see Fig. 4.2)

$$x = x(t,s), \quad y = y(t,s) \quad \text{and} \quad z = z(t,s), \tag{4.42}$$

where is necessary to ask for

$$\frac{\partial(x,y)}{\partial(s,t)} \neq 0 \tag{4.43}$$

to be able to have the inverse relations

$$s = s(x,y) \tag{4.44a}$$

and

$$t = t(x,y). \tag{4.44b}$$

Let us assume that $z = (x,y)$ is twice differentiable and is the integral surface of

$$F(x,y,z,p,q) = 0. \tag{4.45}$$

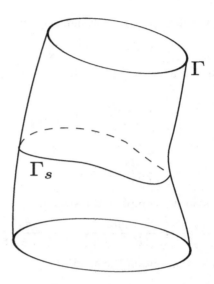

Fig. 4.2   Curve of initial conditions crossing the characteristics.

Doing the differentiation of Eq. (4.45) with respect to x and y, we have

$$F_x + pF_z + F_p \frac{\partial p}{\partial x} + F_q \frac{\partial q}{\partial x} = 0 \qquad (4.46\text{a})$$

and

$$F_y + qF_z + F_p \frac{\partial p}{\partial y} + F_q \frac{\partial q}{\partial y} = 0. \qquad (4.46\text{b})$$

Now, because of equality

$$\frac{\partial q}{\partial x} = \frac{\partial p}{\partial y},$$

Eq. (4.46) can be written as

$$F_x + pF_z + F_p \frac{\partial p}{\partial x} + F_q \frac{\partial p}{\partial y} = 0 \qquad (4.47\text{a})$$

and

$$F_y + qF_z + F_p \frac{\partial q}{\partial x} + F_q \frac{\partial q}{\partial y} = 0, \qquad (4.47\text{b})$$

which represent linear partial differential equation for p and q. The equation for the characteristic of the system (4.47), taking p and q as unknown, are (see (2.16))

$$\frac{dx}{F_p} = \frac{dy}{F_q} = \frac{dp}{-(F_x + pF_z)} = \frac{dq}{-(F_y + qF_z)} = dt, \qquad (4.48)$$

and because z is related with p and q through the expression

$$dz = pdx + qdy,$$

then along one of the characteristic of Eq. (4.48), this expression will given by

$$\frac{dz}{dt} = p\frac{dx}{dt} + q\frac{dy}{dt} = pF_p + qF_q,$$

or

$$\frac{dz}{pF_p + qF_q} = dt. \tag{4.49}$$

With this last expression, we complete the system

$$\frac{dx}{F_p} = \frac{dy}{F_q} = \frac{dz}{pF_p + qF_q} = \frac{dp}{-(F_x + pF_z)} = \frac{dq}{-(F_y + qF_z)} = dt. \tag{4.50}$$

From these equation we can obtain the solution

$$x = x(t), \quad y = y(t), \quad z = z(t), \quad p = p(t) \quad \text{and} \quad q = q(t),$$

that is, we can find the characteristics

$$x = x(t), \quad y = y(t), \quad \text{and} \quad z = z(t).$$

However, one can not guarantee uniqueness of these equations since one has only the initial data $x_0(s)$, $y_0(s)$ and $z_0(s)$, one still needs the conditions $p_0(s)$ and $q_0(s)$. To obtain these conditions, let us see first the behavior of the integral surface of Eq. (4.45) along the characteristic curves. Calculating the total variation of Eq. (4.45) along an integral curve of the system (4.50),

$$\frac{dF}{dt} = F_x\frac{dx}{dt} + F_y\frac{dy}{dt} + F_z\frac{dz}{dt} + F_p\frac{dp}{dt} + F_q\frac{dq}{dt},$$

and using Eq. (4.50), we have

$$\frac{dF}{dt} = F_xF_p + F_yF_q + F_z[pF_p + qF_q] - F_p[F_x + pF_y] - F_q[F_y + qF_z] = 0$$

then, F is constant along an integral curve of the system (4.50), that is

$$F(x, y, z, p, q) = F(x_0, y_0, z_0, p_0, q_0), \tag{4.51}$$

and if Eq. (4.50) is satisfied, we must choose the initial values $x_0(s), y_0(s), z_0(s), p_0(s)$ and $q_0(s)$ such that

$$F(x_0, y_0, z_0, p_0, q_0) = 0. \tag{4.52}$$

This expression give us the first condition to find $p_0(s)$ and $q_0(s)$. To find the other condition, let us assume that $q_0(s)$ is given. Then, integrating the system (4.50) such that Eq. (4.52) is satisfied, we obtain

$$x = x(s,t),\ y = y(s,t),\ z = z(s,t),\ p = p(s,t),\ q = q(s,t). \qquad (4.53\text{a})$$

Fixing s, we obtain one the characteristic curves

$$x = x(s,t), \quad y = y(s,t), \quad z = z(s,t), \qquad (4.53\text{b})$$

and taking the variation of s, we obtain a surface. At any point of this surface, Eq. (4.53) and Eq. (4.45) satisfied. We still need to verify that

$$p = \frac{\partial z}{\partial x} \quad \text{and} \quad q = \frac{\partial z}{\partial y}$$

that is, the following expression must be satisfied

$$dz = p\,dx + q\,dy \qquad (4.54)$$

or, in terms of the variable s and t

$$dz = p\left(\frac{\partial x}{\partial s}ds + \frac{\partial x}{\partial t}dt\right) + q\left(\frac{\partial y}{\partial s}ds + \frac{\partial y}{\partial t}dt\right) = \frac{\partial z}{\partial s}ds + \frac{\partial z}{\partial t}dt$$

which is equivalent to the following conditions

$$p\frac{\partial x}{\partial s} + q\frac{\partial y}{\partial s} = \frac{\partial z}{\partial s} \qquad (4.55)$$

and

$$p\frac{\partial x}{\partial t} + q\frac{\partial y}{\partial t} = \frac{\partial z}{\partial t}. \qquad (4.56)$$

The Eq. (4.56) is an identity since in Eq. (4.50) we already imposed the condition (4.54) along the characteristics. However, Eq. (4.55) is not guaranteed by any expression. Let us find the conditions to satisfy Eq. (4.55). Defining the function U as

$$U = p\frac{\partial x}{\partial s} + q\frac{\partial y}{\partial s} - \frac{\partial z}{\partial s} \qquad (4.57)$$

and differentiating with respect to t, we have

$$\frac{\partial U}{\partial t} = \frac{\partial p}{\partial t}\frac{\partial x}{\partial s} + p\frac{\partial^2 x}{\partial t\partial s} + \frac{\partial q}{\partial t}\frac{\partial y}{\partial s} + q\frac{\partial^2 y}{\partial t\partial s} - \frac{\partial^2 z}{\partial t\partial s}. \qquad (4.58)$$

Now, doing the differentiation of Eq. (4.56) with respect to s, we get

$$\frac{\partial p}{\partial s}\frac{\partial x}{\partial t} + p\frac{\partial^2 x}{\partial s\partial t} + \frac{\partial q}{\partial s}\frac{\partial y}{\partial t} + q\frac{\partial^2 y}{\partial s\partial t} - \frac{\partial^2 z}{\partial s\partial t} = 0$$

which can be used in Eq. (4.58) to obtain

$$\frac{\partial U}{\partial t} = \frac{\partial p}{\partial t}\frac{\partial x}{\partial s} + \frac{\partial q}{\partial t}\frac{\partial y}{\partial s} - \frac{\partial p}{\partial s}\frac{\partial x}{\partial t} - \frac{\partial q}{\partial s}\frac{\partial y}{\partial t}, \tag{4.59}$$

where we have assumed that the characteristic (4.53b) are twice continuously differentiable. Eq. (4.59) along with Eq. (4.50) takes the form

$$\frac{\partial U}{\partial t} = -[F_x + pF_z]\frac{\partial x}{\partial s} - [F_y + qF_z]\frac{\partial y}{\partial s} - F_p\frac{\partial p}{\partial s} - F_q\frac{\partial q}{\partial s}.$$

Rearranging terms in this equation and adding and subtracting $F_z\partial z/\partial s$, it follows that

$$\frac{\partial U}{\partial t} = -\left[F_x\frac{\partial x}{\partial s} + F_z\frac{\partial z}{\partial s} + F_p\frac{\partial p}{\partial s} + F_q\frac{\partial q}{\partial s}\right] - F_z\left[p\frac{\partial x}{\partial s} + q\frac{\partial y}{\partial s} - \frac{\partial z}{\partial s}\right]$$

that is

$$\frac{\partial U}{\partial t} = -\frac{\partial F}{\partial s} - F_z U.$$

But, since Eq. (4.1) is satisfied, it follows that

$$\frac{\partial U}{\partial t} = -F_z U$$

which has the solution

$$U(s,t) = U_0(s)\exp\left(-\int_0^t F_z dt\right). \tag{4.60}$$

If the initial conditions are selected such that $U_0 = 0$, we have that $U \equiv 0$ for any t. Therefore, from the definition (4.57), Eq. (4.55) is satisfied. Consequently, Eq. (4.54) is also satisfied. The initial conditions for the functions p and q, $p_0(s)$ and $q_0(s)$, are deduced from

$$F(x_0, y_0, z_0, p_0, q_0) = 0 \tag{4.61a}$$

and

$$p_0\frac{\partial x_0}{\partial s} + q_0\frac{\partial y_0}{\partial s} = \frac{\partial z_0}{\partial s}. \tag{4.61b}$$

Example 4.8. Find the integral curve of equation

$$z = \left(\frac{\partial z}{\partial x}\right)\left(\frac{\partial z}{\partial y}\right) \tag{4.62}$$

passing through the line $x = 1$, $z = y$.

Let us find first the parametric expressions for the initial curve and the initial values for p and q. The initial parametric curve Γ_s is given by

$$x_0(s) = 1, \quad y_0(s) = s, \quad z_0(s) = s.$$

Using Eq. (4.62) we have

$$s = p_0 q_0$$

and

$$q_0 = 1.$$

The complete initial conditions are

$$x_0(s) = 1, \quad y_0(s) = s, \quad z_0(s) = s, \quad p_0(s) = s \quad \text{and} \quad q_0(s) = 1.$$

Our function F is given by

$$F = pq - z,$$

then the system (4.50) is written as

$$\frac{dx}{q} = \frac{dy}{p} = \frac{dz}{2pq} = \frac{dp}{p} = \frac{dq}{q} = dt,$$

from which the following solution are gotten

$$p = c_1 e^t,$$
$$q = c_2 e^t,$$
$$x = c_2 e^t + d_1,$$
$$y = c_1 e^t + d_2$$

and

$$z = c_1 c_2 e^{2t} + d_3,$$

where c_1, c_2, c_3, d_1, d_2 and d_3 are constant. Using the above initial conditions these constant are determined, obtaining the solution

$$p(s,t) = se^t,$$
$$q(s,t) = e^t,$$
$$x(s,t) = e^t,$$
$$y(s,t) = se^t$$

and

$$z(s,t) = se^{2t}.$$

Eq. (4.43) is readily satisfied

$$\frac{\partial(x,y)}{\partial(s,t)} = -e^t,$$

and the inverse relations (4.44) are

$$s = \frac{x}{y},$$

and

$$t = \log x.$$

Thus, the integral surface satisfying the initial conditions is given by

$$z = xy.$$

Note that this solution is the same as that of Example (4.7), Eq. (4.40b), which was obtained using a different method. The use of particular method depends on the skill of the individual.

Let us make the following two observations:

Observation 1. The special cases (a), (b), (c) and (d) presented at the beginning of this chapter can be deduced from the Lagrange-Charpit method by solving Eq. (4.27) with the corresponding form F of the nonlinear PDEFO. This is left as exercises at the end of this chapter.

Observation 2. One must note that in the case of having a quasi-linear (or linear) PDEFO defined in $\mathbb{R}^2$ as Eq. (2.15), this one can be expressed as

$$F = pP(x, y, z) + qQ(x, y, z) - R(x, y, z) = 0,$$

having the following differentiations

$$F_x = pP_x + qQ_x - R_x, \quad F_y = pP_y + qQ_y - R_y, \quad F_z = pP_z + qQ_z - R_z,$$
$$F_p = P, \quad F_q = Q.$$

Therefore, the equations for the characteristics Eq. (4.50) are written for this case as

$$\frac{dx}{P(x,y,z)} = \frac{dy}{Q(x,y,z)} = \frac{dz}{R(x,y,z)} = \frac{-dp}{pP_x + qQ_x - R_x + p(pP_z + qQ_z - R_z)}$$
$$= \frac{-dq}{pP_y + qQ_y - R_y + q(pP_z + qQ_z - R_z)}.$$

The first three terms on the left side are decoupled from the other terms and defines the characteristic curves for this type of PDEFO, Eq. (2.16a). Thus, the projection of the characteristics in the space (x, y, z) of this quasi-linear PDEFO brings about the characteristic curves studied before for these type of equations defined in $\mathbb{R}^2$.

4.2 Non-Linear PDEFO for Functions Defined in $\mathbb{R}^n$

The PDEFO of functions defined in n-dimensional space $(x_1, ..., x_n)$ have the form

$$F(\mathbf{x}, z, \mathbf{p}) = 0, \qquad (4.63)$$

where $\mathbf{x}$ and $\mathbf{p}$ are n-dimensional vectors defined as

$$\mathbf{x} = (x_1, ..., x_n) \qquad (4.64\text{a})$$

and

$$\mathbf{p} = (p_1, ..., p_n), \qquad (4.64\text{b})$$

where p_i is given by

$$p_i = \frac{\partial z}{\partial x_i} \quad i = 1, ..., n \qquad (4.65)$$

and $z = z(\mathbf{x})$ is a n-dimensional surface. It is clear that the Lagrange-Charpit idea for solving PDEFO defined in $\mathbb{R}^n$ does not work any more for $n > 2$ since the integrability condition of the Pfaffian equation, $\sum_{i=1}^{n} p_i dx_i - dz = 0$, would imply the coupling of proposed functions $U_i(\mathbf{x}, z, \mathbf{p}) = a_i$ for obtaining $p_i = p_i(\mathbf{x}, z, \mathbf{a})$. However, one will see that the Cauchy initial conditions approach will work, in principle, for arbitrary dimensional space $\mathbb{R}^n$. Cauchy's problem for Eq. (4.63) consists of determine the n-dimensional surface passing through the $n-1$ dimensional surface Γ_s given by

$$z_0 = z_0(\mathbf{s}), \quad x_{i0} = x_{i0}(\mathbf{s}) \qquad (4.66\text{a})$$

where $\mathbf{s}$ is a $n-1$ dimensional vector

$$\mathbf{s} = (s_1, ..., s_{n-1}). \qquad (4.66\text{b})$$

What we have done for two-dimensional space is immediately generalized to this n-dimensional space. Making the differentiation of Eq. (4.65) whit respect to the variable x_i,

$$F_{x_i} + p_i F_z + \sum_{l=1}^{n} F_{p_l} \frac{\partial p_l}{\partial x_i} = 0 \quad i = 1, ..., n, \qquad (4.67)$$

and assuming that the function z is twice continuously differentiable,

$$\frac{\partial p_l}{\partial x_i} = \frac{\partial^2 z}{\partial x_i \partial x_l} = \frac{\partial p_i}{\partial x_l} \quad i, l = 1, ..., n,$$

Eq. (4.67) can be written as

$$\sum_{l=1}^{n} F_{p_l} \frac{\partial p_i}{\partial x_l} = -(F_{x_i} + p_i F_z) \quad i = 1, ..., n. \tag{4.68}$$

This represents a partial differential equation for p_i $(i = 1, ..., n)$, and the equations for its characteristics curves are given by

$$\frac{dx_1}{F_{p_1}} = ... = \frac{dx_n}{F_{p_n}} = \frac{dp_1}{-(F_{x_i} + p_i F_z)} = ... = \frac{dp_n}{-(F_{x_n} + p_n F_z)} = dt \tag{4.69}$$

which can be completed by using the fact that along the characteristic curves, we must have

$$\frac{dz}{dt} = \sum_{l=1}^{n} p_l \frac{dx_l}{dt} = \sum_{l=1}^{n} p_l F_{p_l}. \tag{4.70}$$

Then, we can written the equation system for the characteristic as

$$\frac{dx_1}{F_{p_1}} = ... = \frac{dx_n}{F_{p_n}} = \frac{dz}{\displaystyle\sum_{l=1}^{n} p_l F_{p_l}} = \frac{dp_1}{-(F_{x_1} + p_1 F_z)} = \frac{dp_n}{-(F_{x_n} + p_n F_z)} = dt. \tag{4.71}$$

Thus, assuming we know the initial values for p_i,

$$p_{i0} = p_{i0}(\mathbf{s}) \quad i = 1, ..., n, \tag{4.72}$$

and integrating the system (4.71), we can obtain the solution

$$x_i = x_i(\mathbf{s}, t) \quad i = 1, ..., n, \tag{4.73a}$$

$$z = z(\mathbf{s}, t), \tag{4.73b}$$

and

$$p_i = p_i(\mathbf{s}, t) \quad i = 1, ..., n \tag{4.73c}$$

which satisfy the initial conditions (4.66a) and (4.72). For a fixed vector $\mathbf{s}$, the solution (4.73a) and (4.73b) determine a curve in the $(n+1)$-dimensional space $(\mathbf{x}, z)$, called characteristic. As the vector $\mathbf{s}$ changes, we generated a n-dimensional surface passing through the initial $(n-1)$-dimensional surface Γ_s. In order for the inverse relation (4.73a) to be satisfied, we have to ask for the Jacobian of the transformation to be different form zero (see reference of T.M. Apostol),

$$\frac{\partial(\mathbf{x})}{\partial(\mathbf{s}, t)} = det \begin{pmatrix} \dfrac{\partial x_1}{\partial s_1} & \cdots & \dfrac{\partial x_1}{\partial s_{n-1}} & \dfrac{\partial x_1}{\partial t} \\ \vdots & \cdots & \vdots & \vdots \\ \dfrac{\partial x_n}{\partial s_1} & \cdots & \dfrac{\partial x_n}{\partial s_{n-1}} & \dfrac{\partial x_n}{\partial t} \end{pmatrix} \neq 0. \tag{4.73d}$$

We will show that with a suitable selection of values for p_{i0} $(i = 1, ..., n)$, the set of points determined by the characteristics (4.73) form the n-dimensional integral surface of Eq. (4.63). These values must satisfy

$$F(\mathbf{x}(\mathbf{s}, t), z(\mathbf{s}, t), \mathbf{p}(\mathbf{s}, t)) = 0 \qquad (4.74)$$

and

$$p_i = \frac{\partial z}{\partial x_i} \quad i = 1, ..., n \qquad (4.75a)$$

or equivalently

$$dz = \sum_{l=1}^{n} p_l dx_l. \qquad (4.75b)$$

Differentiating the function F with respect to t, we have

$$\frac{dF}{dt} = \sum_{i=1}^{n} F_{x_i} \frac{dx_i}{dt} + F_z \frac{dz}{dt} + \sum_{i=1}^{n} F_{p_i} \frac{dp_i}{dt},$$

and considering this equation along the integral curves of the system (4.71), we get

$$\frac{dF}{dt} = \sum_{i=1}^{n} F_{x_i} F_{p_i} + F_z \sum_{i=1}^{n} p_i F_{p_i} - \sum_{i=1}^{n} F_{p_i} (F_{x_i} + p_i F_z) = 0.$$

Consequently, the function F is constant along these curves, that is

$$F(\mathbf{x}, z, \mathbf{p}) = F(\mathbf{x}_0, z_0, \mathbf{p}_0). \qquad (4.76)$$

Thus, the values of p_{i_0} $(i = 1, ..., n)$ must be such that

$$F(x_{1_0}, ..., x_{n_0}, z_0, p_{1_0}, ..., p_{n_0}) = 0. \qquad (4.77)$$

We still need to show that the expression

$$dz = \sum_{i=1}^{n} p_i dx_i \qquad (4.78)$$

is satisfied for every t and s_j $(j = 1, ..., n - 1)$. Expressing Eq. (4.78) in terms these parameters, it follows that

$$dz = \frac{\partial z}{\partial t} dt + \sum_{j=1}^{n-1} \frac{\partial z}{\partial s_j} ds_j = \sum_{i=1}^{n} p_i \left[\frac{\partial x_i}{\partial t} dt + \sum_{j=1}^{n} \frac{\partial x_i}{\partial s_j} ds_j \right],$$

and equaling coefficients of the independent variables, we obtain two conditions

$$\frac{\partial z}{\partial t} - \sum_{i=1}^{n} p_i \frac{\partial x_i}{\partial t} = 0 \qquad (4.79a)$$

and

$$\frac{\partial z}{\partial s_j} - \sum_{i=1}^{n} p_i \frac{\partial x_i}{\partial s_j} = 0. \tag{4.79b}$$

Eq. (4.79a) represents an identity along the integral curves of the system (4.71) since on these curves we have

$$\frac{\partial z}{\partial t} = \sum_{i=1}^{n} p_i F_{p_i} \quad \text{and} \quad \frac{\partial x_i}{\partial t} = F_{p_i} \quad i = 1, ..., n.$$

However, Eq. (4.79b) will be true only for certain values of p_{i0} $(i = 1, ..., n)$. To see this, define the function U_j as

$$U_j = \frac{\partial z}{\partial s_j} - \sum_{i=1}^{n} p_i \frac{\partial x_i}{\partial s_j} \quad j = 1, ..., n-1, \tag{4.80}$$

and carrying out the differentiation of Eq. (4.79a) and Eq. (4.79b) with respect to parameter t, we get the following two equations

$$\frac{\partial U_j}{\partial t} = \frac{\partial^2 z}{\partial t \partial s_j} - \sum_{i=1}^{n} p_i \frac{\partial^2 x_i}{\partial t \partial s_j} - \sum_{i=1}^{n} \frac{\partial p_i}{\partial t} \frac{\partial x_i}{\partial s_j} \tag{4.81a}$$

and

$$\frac{\partial^2 z}{\partial t \partial s_j} - \sum_{i=1}^{n} p_i \frac{\partial^2 x_i}{\partial t \partial s_j} - \sum_{i=1}^{n} \frac{\partial p_i}{\partial s_j} \frac{\partial x_i}{\partial t} = 0. \tag{4.81b}$$

Using Eq. (4.81b) in Eq. (4.81a) and the assumption that $x, y, z \in C^2(\mathbb{R}^2)$, we obtain for Eq. (4.81a)

$$\frac{\partial U_j}{\partial t} = \sum_{i=1}^{n} \frac{\partial p_i}{\partial s_j} \frac{\partial x_i}{\partial t} - \sum_{i=1}^{n} \frac{\partial p_i}{\partial t} \frac{\partial x_i}{\partial s_j}. \tag{4.82}$$

Evaluating this expression along the integral curves of Eq. (4.71), we have

$$\frac{\partial U_j}{\partial t} = \sum_{i=1}^{n} \frac{\partial p_i}{\partial s_j} F_{p_i} + \sum_{i=1}^{n} (F_{x_i} + p_i F_z) \frac{\partial x_i}{\partial s_j}.$$

Adding and subtracting the term $F_z(\partial z/\partial s_j)$, it follows that

$$\frac{\partial U_j}{\partial t} = \sum_{i=1}^{n} \left[F_{p_i} \frac{\partial p_i}{\partial s_j} + F_{x_i} \frac{\partial x_i}{\partial s_j} \right] + F_z \frac{\partial z}{\partial s_j}$$

$$- F_z \left[\frac{\partial z}{\partial s_j} - \sum_{i=1}^{n} p_i \frac{\partial x_i}{\partial s_j} \right] = \frac{\partial F}{\partial s_j} - F_z U_j.$$

Taking into consideration Eq. (4.79b), it follows that $\partial F/\partial s_j = 0$, and the solution of the above equation is

$$U_j = U_{j0}(\mathbf{s}) \exp\left(-\int_0^t F_z dt\right).$$ (4.83)

In this way, if we want Eq. (4.79b) to be satisfied, we must have that $U_j = 0$ for any t, and to get this it is enough to have $U_{j0} = 0$ which implies from Eq. (4.80) that the p_{i0} $(i = 1, ..., n)$ must be such that

$$\frac{\partial z_0}{\partial s_j} - \sum_{i=1}^n p_{i0} \frac{\partial x_{i0}}{\partial s_j} = 0 \quad j = 1, ..., n-1.$$ (4.84)

Once we have the values p_{i0} for $i = 1, ..., n$ satisfying the Eq. (4.77) and Eq. (4.84), the solution (4.73) is completely determined.

Example 4.9. Find the integral surface of the equation

$$\sum_{i=1}^n \left(\frac{\partial z}{\partial x_i}\right)^2 = 4$$

passing through the $(n-1)$-dimensional surface Γ_s defined by

$$x_{n0} = 1, \quad z = \sum_{i=1}^{n-1} x_i.$$

Let us find first the parametric form of this initial conditions. Our function F is

$$F = \sum_{i=1}^n p_i^2 - 4 = 0,$$

and the $(n-1)$-dimensional surface Γ_s is given by

$$x_{n0} = 1, \quad z_0 = \sum_{i=1}^{n-1} s_i, \quad x_{i0} = s_i \quad i = 1, ..., n-1.$$

Eq. (4.84) brings about the following relations

$$\frac{\partial z_0}{\partial s_j} - \sum_{i=1}^n p_{i0} \frac{\partial x_{i0}}{\partial s_j} = 1 - p_{j0} = 0 \quad j = 1, ..., n-1$$

and

$$p_{n0} = \pm\sqrt{5-n}.$$

Now, the system (4.71) has the following form

$$\frac{dx_i}{2p_i} = ... = \frac{dx_n}{2p_n} = \frac{dz}{2\sum_{i=1}^n p_i^2} = \frac{dp_1}{0} = ... = \frac{dp_n}{0} = dt$$

which has the solutions

$$p_i = c_i \qquad i = 1, ..., n,$$
$$x_i = 2c_i t + d_i \qquad i = 1, ..., n,$$

and

$$z = 2 \left(\sum_{i=1}^{n} c_i^2 \right) t + d_0.$$

Using the initial conditions in these expressions, it follows that

$$c_n = \pm\sqrt{5 - n} \quad c_i = 1 \quad i = 1, ..., n - 1,$$
$$d_n = 1 \qquad d_i = s_i \quad i = 1, ..., n - 1,$$

and

$$d_0 = \sum_{i=1}^{n} s_i.$$

So, the solution of problem is given by

$$x_i(\mathbf{s}, t) = 2t + s_i \quad i = 1, ..., n - 1, \tag{4.85a}$$
$$x_n(\mathbf{s}, t) = \pm 2\sqrt{5 - n}\,t + 1, \tag{4.85b}$$

and

$$z(\mathbf{s}, t) = 8t + \sum_{i=1}^{n-1} s_i. \tag{4.85c}$$

From Eq. (4.85a) and Eq. (4.85b), we see that

$$\frac{\partial(\mathbf{x})}{\partial(\mathbf{s}, t)} = 1,$$

and then the inverse relations can be given which are written as

$$s_i = x_i \pm \frac{x_n - 1}{\sqrt{5 - n}} \quad i = 1, ..., n - 1,$$

and

$$t = \pm \frac{x_n - 1}{\sqrt{5 - n}}.$$

Thus, using these relations and Eq. (4.85c), the integral surfaces can be integrated as

$$z(\mathbf{x}) = \pm(x_n - 1)\sqrt{5 - n} + \sum_{i=1}^{n-1} x_i.$$

Note that real solutions are obtained for this problem for the dimension of the space lower or equal to five, $n \le 5$.

4.3   Problems

4.1   Find the envelope to the family of surfaces defined by

$$z = ax + a^{-1}y - (b - x)^2.$$

4.2   Find the integral surfaces of the equation

$$\left(\frac{\partial z}{\partial x}\right)^2 - \left(\frac{\partial z}{\partial x}\right)\left(\frac{\partial z}{\partial y}\right)^2 = 4.$$

Suggestion sees the Example 4.2.

4.3   Find the solution of the following equation

$$e^{2y}\left(\frac{\partial z}{\partial x}\right)^3 - x^{-3}\left(\frac{\partial z}{\partial y}\right)^{1/3} .$$

Suggestion sees the Example 4.3.

4.4   Find the solutions of the following equation

$$\left(\frac{\partial z}{\partial x}\right)\left(\frac{\partial z}{\partial y}\right) = 9z^2.$$

Suggestion sees the Example 4.2.

4.5   Find the integral surfaces of the equation

$$xyz = x^2y\frac{\partial z}{\partial x} + y^2x\frac{\partial z}{\partial y} + xy\left[\left(\frac{\partial z}{\partial x}\right)^2 - \left(\frac{\partial z}{\partial y}\right)^2\right].$$

Suggestion sees the Example 4.5.

4.6   Find the integral surfaces of the equation

$$yz\frac{\partial z}{\partial x} - \left(\frac{\partial z}{\partial y}\right)^2 = 0.$$

Suggestion sees the Example 4.6.

4.7   Find te integral surfaces of the equation

$$z = x\frac{\partial z}{\partial x} + y\frac{\partial z}{\partial y} + \frac{1}{4}\frac{\partial z}{\partial x}\frac{\partial z}{\partial y}$$

that passes through the curve $y = 0$, $z = x^2$.

4.8 Find the solution of the problem 4.3, i.e. find the integral surface (using cauchy's method) of the equation

$$F = xp + yq + \frac{1}{4}pq - z = 0$$

which passes through the curve $y = a$, $z = x^2$.

4.9 Find the integral surfaces of the equation

$$\sum_{i=1}^{n} \left(\frac{\partial z}{\partial x_i} \right)^2 - z = 4$$

which passes through the $(n-1)$-dimensional surface Γ_s determined by

$$x_{n0} = 1, \quad z = \sum_{i=1}^{n-1} x_i.$$

4.10 Find the solution of the PDEFO defined in $\mathbb{R}^2$ by

$$z = x\frac{\partial z}{\partial x} + y\frac{\partial z}{\partial y} + \frac{1}{4}\frac{\partial z}{\partial x}\frac{\partial z}{\partial y}$$

such that $z(x,0) = x^2$.

4.11 Demonstrate that the solution of the PDEFO

$$z = \frac{\partial z}{\partial x}\frac{\partial z}{\partial y}$$

which passes through the initial curve $\Gamma_o : \{x = 1, z = y$ is given by $z(x,y) = xy$.

4.12 Find the envelope solution of the PDEFO

$$yz\left(\frac{\partial z}{\partial x}\right) - \left(\frac{\partial z}{\partial y}\right)^2 = 0.$$

4.13 Use the Lagrange-Charpit method to find the integral surface of the non linear PDEFO defined by
i) $F(p,q) = 0$.
ii) $F(x,y,p,q) = f_1(x,p) - f_2(y,q) = 0$.
iii) $F(z,p,q) = 0$.
iv) $F(x,y,z,p,q) = px + qy + f(p,q) - z = 0$.

4.4 References

4.1 L. Elsgoltz, *Ecuaciones Diferenciales y Cálculo Variacional*, Ediciones Cultura Popular, S.A., 1983. Chap. 5.

4.2 N. Piskunov, *Differential and Integral Calculus*, MIR Publisher, 1969. Chap. 13.

4.3 T.M. Apostol, *Mathematical Analysis*, Addison-Wesley Publishing Co., 1965. Pages 146-148.

4.4 R. Courant and D. Hilbert, *Methods of Mathematical Physics*, John Weley & Sons, N.Y., 1962.

Chapter 5

Physical Applications II

In this chapter we will illustrate the use of the PDEFO in some problems of Physics. We will not try to explain the full physical concepts or to develop the full physical consequences of solution of the equations, this is not the intention of this book, but we will give the proper references where the full physical understanding of the solution can be found.

5.1 Motion of a Classical Particle

Following the ideas of Sec. 1.6, we will present some solutions for the non-linear PDEFO appearing in that chapter.

5.1.1 Hamilton-Jacobi Equation for a One-Dimensional Harmonic Oscillator

When we solved exercise 6 of Chapter 1, Section 6, we found that the Hamiltonian for the one dimensional harmonic oscillator was given by the expression

$$H = \frac{p^2}{2m} + \frac{1}{2}kq^2 \,, \tag{5.1}$$

where m is the mass of the particle and k is the spring constant. With this Hamiltonian, the Hamilton-Jacobi equation is given, using (1.75), by

$$\frac{1}{2m}\left(\frac{\partial S}{\partial q}\right)^2 + \frac{1}{2}kq^2 + \frac{\partial S}{\partial t} = 0 \,. \tag{5.2}$$

According to Eq. (4.1), the function F is given by

$$F(t, q, S, S_t, S_q) = \frac{1}{2m}S_q^2 + \frac{1}{2}kq^2 + S_t = 0 \,. \tag{5.3}$$

So, the partial differentiations are

$$F_t = 0, \quad F_q = kq, \quad F_s = 0, \quad F_{S_t} = 1, \quad F_{S_q} = S_q/m,$$

and using these in Eq. (4.50), we have the following system

$$dt = \frac{dq}{S_q/m} = \frac{dS}{S_t + S_q^2/m} = \frac{dS_t}{0} = \frac{dS_q}{-kq}.$$

From the second and last term, we get the equation

$$\frac{dq}{S_q/m} = \frac{dS_q}{-kq}$$

and solving this one, we obtain the function U as

$$U = \frac{S_q^2}{2m} + \frac{1}{2}kq^2 = a, \tag{5.4}$$

where a is a constant of integration. From Eq. (5.3) and (5.4), we get

$$\frac{\partial(F, U)}{\partial(S_t, S_q)} = S_q/m,$$

then for every value such that $S_q \neq 0$, we can have

$$S_t = -a \tag{5.5a}$$

and

$$S_q = \pm\sqrt{km}\left[\frac{2a}{k} - q^2\right]^{1/2} \tag{5.5b}$$

and integrating the expression

$$ds = S_t dt + S_q dq,$$

we obtain the solution of Eq. (5.2) as

$$S = -at \pm \sqrt{mk}\int\left[\frac{2a}{k} - q^2\right]^{1/2} dq + b. \tag{5.6}$$

The trajectory of the particle can be deduced from Eq. (1.78)

$$\beta = \frac{\partial S}{\partial a}.$$

Using Eq. (5.6), we obtain

$$\beta = -t \pm \sqrt{m/k} \int \frac{dq}{[2a/k - q^2]^{1/2}},$$

and after rearranging terms and making the integration, the solution can be written as

$$q = \sqrt{2a/k} \cos \omega(t + \beta),　\qquad (5.7a)$$

where a represents the energy of the particles and ω is the angular frequency of oscillation given as

$$\omega = \sqrt{k/m}.　\qquad (5.7b)$$

The constants a and β are determined by the initial conditions of the problem, (q_0, p_0).

5.1.2 *The Lagrangian Obtained Directly from Hamiltonian*

If we have the Hamiltonian of a system in some way, we saw in Chapter 1, Section 5 that it is possible to obtain directly the Lagrangian of the system by integrating the nonlinear PDEFO given by Eq. (1.67). Suppose that we have the one dimensional dissipative system of Example 6, Chapter 1, Section 5, whose Hamiltonian is given by

$$H = \frac{p^2}{2m} \exp(-2\alpha q/m),　\qquad (5.8)$$

where m is the mass of the particle and α is the friction constant characteristic of the medium. The form of Eq. (1.67) for this case is written as

$$\frac{\exp(-2\alpha x/m)}{2} \left(\frac{\partial L}{\partial v}\right)^2 - v\frac{\partial L}{\partial v} + L = 0,　\qquad (5.9)$$

where we have changed the notation ($q \to x$, and $\dot{q} \to v$) for convenience. Defining now $\tilde{p}$ and $\tilde{q}$ as

$$\tilde{p} = \frac{\partial L}{\partial x} \qquad \text{and} \qquad \tilde{q} = \frac{\partial L}{\partial v},　\qquad (5.10)$$

the function F (see Eq. (4.1)), is given by

$$F(x, v, L, \tilde{p}, \tilde{q}) = \frac{\exp(-2\alpha x/m)}{2m}\tilde{q}^2 - v\tilde{q} + L = 0,　\qquad (5.11)$$

and the partial differentiations are

$$F_x = -\frac{\alpha \exp(-2\alpha x/m)}{m^2}\tilde{q}^2\,,$$

$$F_v = -\tilde{q}\,,$$

$$F_L = 1\,,$$

$$F_{\tilde{p}} = 0\,,$$

and

$$F_{\tilde{q}} = \frac{\exp(-2\alpha x/m)}{m}\tilde{q} - v\,.$$

Thus the system (4.50) is written as

$$\frac{dx}{0} = \frac{dv}{\dfrac{\exp(-2\alpha x/m)}{m}\tilde{q} - v} = \frac{dL}{\dfrac{\exp(-2\alpha x/m)}{m}\tilde{q}^2 - v\tilde{q}}$$

$$= \frac{d\tilde{p}}{\alpha \exp(-2\alpha x/m)\tilde{q}^2 - \tilde{p}} = \frac{d\tilde{q}}{0}\,.$$

From the last term of this system, we can immediately obtain the function $\tilde{q}$ given by

$$\tilde{q} = a\,,$$

where a is a constant integration. Using this in Eq. (5.11), we get a one-parameter family of solutions given by

$$\Phi(x, v, L, a) = -\frac{\exp(-2\alpha x/m)}{2m}a^2 + va - L = 0\,, \qquad (5.12)$$

the envelope of the solution is given by (5.1) itself and the relation

$$\frac{\partial \Phi}{\partial a} = -\frac{\exp(-2\alpha x/m)}{m}a + v = 0\,. \qquad (5.13)$$

The relation (5.13) give us

$$a = mve^{2\alpha x/m} \qquad (5.14)$$

and using this in Eq. (5.12), we finally obtain

$$L = \frac{1}{2}mv^2 e^{2\alpha x/m}\,, \qquad (5.15)$$

which is the same we have in expression (1.83). Of course, this same expression could be obtained by using Eq. (1.65a) to get the relation between v and p, and then using Eq. (1.64).

5.1.3 *Relativistic Particle Moving in a Coulomb Field*

The Hamiltonian for a relativistic motion of a charged particle in a electromagnetic field, characterized by the vector potential **A** and the scalar potential ϕ, is given (see reference L.C. Landau and E.M. Lifshitz) by

$$H = \left[m^2 c^4 + c^2 (\mathbf{P} - \frac{e}{c}\mathbf{A})^2 \right]^{1/2} + e\phi, \qquad (5.16)$$

where m is the mass of the particle, c is the speed of light, e is the charge of the particle which is an integer multiple of the electron charge (1.602×10^{-19} Coulomb) because the charge is quantized, and $\mathbf{P}$ is the relativistic momentum vector of the particle. Using this Hamiltonian in the expression (1.76), squaring the expression obtained, and rearranging terms, we obtain the Hamilton-Jacobi equation

$$\left(\nabla S - \frac{e}{c}\mathbf{A} \right)^2 - \frac{1}{c^2} \left(\frac{\partial S}{\partial t} + e\phi \right)^2 + \kappa^2 = 0 , \qquad (5.17)$$

where we have define $\kappa = mc$. The Coulomb field produced by a point charge e' is characterized by the following expression for the potential **A** and ϕ

$$\mathbf{A} = 0 \qquad \text{and} \qquad \phi = e'/r , \qquad (5.18)$$

where r is the distance from the charge e' to e in spherical coordinates. In these coordinates, Eq. (5.17) has the following expression for our problem

$$\left(\frac{\partial S}{\partial r} \right)^2 + \frac{1}{r^2} \left(\frac{\partial S}{\partial \theta} \right)^2 + \frac{1}{r^2 \sin^2 \theta} \left(\frac{\partial S}{\partial \phi} \right)^2$$
$$- \frac{1}{c^2} \left(\frac{\partial S}{\partial t} + \frac{\alpha}{r} \right)^2 + \kappa^2 = 0 \qquad (5.19)$$

where α is given by $\alpha = ee'$. Defining the variables p_1, p_2, p_3 and p_4 as

$$p_1 = \frac{\partial S}{\partial r}, \quad p_2 = \frac{\partial S}{\partial \theta}, \quad p_3 = \frac{\partial S}{\partial \phi} \quad \text{and} \quad p_4 = \frac{\partial S}{\partial t},$$

the function F (Eq. (4.63)) is given as

$$F(r, \theta, \phi, t, S, p_1, p_2, p_3, p_4) = p_1 + \frac{p_2^2}{r^2} + \frac{p_3^2}{r^2 \sin^2 \theta}$$
$$- \frac{1}{c^2} \left(p_4 + \frac{\alpha}{r} \right)^2 + \kappa^2 = 0 . \qquad (5.20)$$

The partial differentiations of this function are

$$F_r = -2\frac{p_2^2}{r^3} - 2\frac{p_3^2}{r^3 \sin^2\theta} + \frac{2}{c^2}\left(p_4 + \frac{\alpha}{r}\right)\frac{\alpha}{r^2}\,,$$

$$F_\theta = -2\frac{p_3^2 \cos\theta}{r^2 \sin^3\theta}\,,$$

$$F_\phi = 0\,,$$

$$F_t = 0\,,$$

$$F_{p_1} = 2p_1\,,$$

$$F_{p_2} = 2p_2/r^2\,,$$

$$F_{p_3} = 2p_3/r^2 \sin^2\theta \quad \text{and}$$

$$F_{p_4} = -2(p_4 + \alpha/r)/c^2\,.$$

Using this differentiations in Eq. (4.71), we obtain the system of equations

$$\frac{dr}{2p_1} = \frac{d\theta}{\dfrac{2p_2}{r^2}} = \frac{d\phi}{2p_3/r^2 sin^2\theta} = \frac{dt}{-2(p_4 + \alpha/r)/c^2}$$

$$= \frac{dS}{2p_1^2 + \dfrac{2p_2^2}{r^2} + \dfrac{2p_3^2}{r^2 \sin^2\theta} - \dfrac{2p_4}{c^2}(p_4 + \alpha/r)}$$

$$= \frac{dp_1}{2\dfrac{p_2^2}{r^3} + 2\dfrac{p_3^2}{r^3 \sin^2\theta} - \dfrac{2}{c^2}(p_4 + \alpha/r)\dfrac{\alpha}{r^2}}$$

$$= \frac{dp_2}{2\dfrac{p_3^2 \cos\theta}{r^2 \sin^3\theta}} = \frac{dp_3}{0} = \frac{dp_4}{0}\,. \tag{5.21}$$

From the last two terms, we have for p_3 and p_4

$$p_3 = a_1 \tag{5.22a}$$

and

$$p_4 = a_2\,. \tag{5.22b}$$

Using Eq. (5.22a) in Eq. (5.21) and the terms containing $d\theta$ and dp_2, we have the equation

$$\frac{d\theta}{\dfrac{2p_2}{r^2}} = \frac{dp_2}{2\dfrac{p_3^2 \cos\theta}{r^2 \sin^3\theta}}\,,$$

whose solution is

$$p_2^2 + \frac{a_1^2}{\sin^2\theta} = a_3^2\,, \tag{5.22c}$$

where a_3 is the constant of integration. Substituting Eq. (5.22) in Eq. (5.21) and taking the terms containing dr and dp_1, we have the equation

$$\frac{dr}{2p_1} = \frac{dp_1}{2\dfrac{a_3^2}{r^3} - \dfrac{2}{c^2}(a_2 + \alpha/r)\dfrac{\alpha}{r^2}}$$

and making the integration, we get

$$p_1^2 = -\frac{a_3^2}{r^2} + \frac{2a_2\alpha}{c^2}\frac{1}{r} + \frac{\alpha^2}{c^2}\frac{1}{r^2} + a_4\,, \tag{5.22d}$$

where a_4 is another constant of integration. Using the expression (5.22) in the differential

$$dS = p_1 dr + p_2 d\theta + p_3 d\phi + p_4 dt$$

and integrating, we get the solution of Eq. (5.19) as

$$S = a_1\phi + a_2 t \pm \int \left[-\frac{a_3^2}{r^2} + \frac{2a_2\alpha}{c^2}\frac{1}{r} + \frac{\alpha^2}{c^2}\frac{1}{r^2} + a_4 \right]^{1/2} dr$$

$$\pm \int [a_3^2 - a_1^2/\sin^2\theta]^{1/2} d\theta + b\,, \tag{5.23}$$

where b is an integration constant. The constant a_4 is not independent of the other ones because by substituting p_1, p_2, p_3 and p_4 in Eq. (5.20), we get

$$a_4 = \frac{a_2^2}{c^2} - \kappa^2\,, \tag{5.24}$$

and we can express Eq. (5.23) then as

$$S = a_1\phi + a_2 t \pm \int \left[\frac{(a_2 + \alpha/r)^2}{c^2} - \frac{a_3^2}{r^2} - \kappa^2 \right]^{1/2} dr$$

$$\pm \int [a_3^2 - a_1^2/\sin^2\theta]^{1/2} d\theta\,, \tag{5.25}$$

where a_1 represents the energy of the particle and a_3 its angular momentum. The trajectory of the particle is determined from relation (1.78)

$$\beta = \frac{\partial S}{\partial a_3}\,. \tag{5.26}$$

The trajectory of the particle is given by the algebraic solution of the expression

$$\beta = \mp \int \frac{a_3 dr}{r^2 \left[\frac{1}{c^2}(a_2 + \alpha/r)^2 - \frac{a_3^2}{r^2} - \kappa^2 \right]^{1/2}} \pm \int \frac{a_3 \sin\theta d\theta}{[a_3^2 \sin^2\theta - a_1^2]^{1/2}}\,. \tag{5.27}$$

A physical discussion of the solution, for the particular case $a_1 = 0$, can be found in the same reference mentioned previously in this subsection.

5.1.4 *Motion of a Test Particle in a Schwarchild's Space*

If in Eq. (5.17), we make $\mathbf{A} = 0$ and $\phi = 0$, the equation can be written as

$$\eta^{\mu\nu}\left(\frac{\partial S}{\partial x^{\mu}}\right)\left(\frac{\partial S}{\partial x^{\nu}}\right) + \kappa^2 = 0\,, \qquad (5.28)$$

where repeated indexes means summation over these (Einstein's convention), in cartesian coordinates, x^{μ} for $\mu = 1, 2, 3, 4$ are the components of the vector $\mathbf{x} = (x, y, z, ct)$ and $\eta^{\mu\nu}$ are the components of the 4×4 matrix

$$(\eta^{\mu\nu}) = \begin{bmatrix} 1 & 0 & 0 & 0 \\ 0 & 1 & 0 & 0 \\ 0 & 0 & 1 & 0 \\ 0 & 0 & 0 & -1 \end{bmatrix} \qquad (5.29)$$

which is the inverse matrix of that one formed with the coefficients of the pseudo-metric in the four dimensional Minskowsky space (flat space). This metric is

$$ds^2 = \eta_{\mu\nu}dx^{\mu}dx^{\nu} = dx^2 + dy^2 + dz^2 - c^2dt^2\,. \qquad (5.30)$$

In this way, Eq. (5.28) is interpreted as the Hamilton-Jacobi equation for a free test particle moving in the flat space defined by the metric (5.30). General relativity is a correction of the Newton gravity theory for strong gravity forces which considers that any mass in space-time, and in fact any form of energy, causes a deformation of our flat space in such a way that the metric becomes

$$ds^2 = g_{\mu\nu}(\mathbf{x})dx^{\mu}dx^{\nu}\,, \qquad (5.31)$$

where the coefficients of our metric $g_{\mu\nu}(\mathbf{x})$ are functions determined by the kind of energy, matter, and the symmetry of this one. These coefficients can be calculated by the Einstein's equation (see reference C. Møller). Thus, we have passed from a pseudo-Euclidean space described by Eq. (5.30) to a pseudo-Riemann space characterized by the metric (5.31). Taking the inverse matrix $(g^{\mu\nu}(\mathbf{x}))$, we can calculate the free motion of a test particle (it is assumed that the deformation caused in the space by the particle is negligible) in this space with the equation

$$g^{\mu\nu}(\mathbf{x})\left(\frac{\partial S}{\partial x^{\mu}}\right)\left(\frac{\partial S}{\partial x^{\nu}}\right) + \kappa^2 = 0\,. \qquad (5.32)$$

A solution of the Einstein equation for a centrally symmetric gravitational field is the Schwarzchild solution, given in spherical coordinates as

$$ds^2 = -\left(1 - \frac{r_g}{r}\right) c^2 dt^2 + r^2 (\sin^2\theta d\phi^2 + d\theta^2)$$
$$+ \left(1 - \frac{r_g}{r}\right)^{-1} dr^2 , \quad (5.33)$$

where r_g is called Schwarzchild's radius and it is given by

$$r_g = \frac{2kM}{c^2} , \quad (5.34)$$

where k is the gravitational constant ($k = 6.67 \times 10^{-8} cm^3/gm\ sec^2$), and M is the mass of the object causing this field. The kind of singularity appearing in the metric (5.33) when $r = r_g$ is called *chart singularity* because it depends on the particular coordinates we are using to describe the metric (in other coordinates, this singularity can disappear). The metric matrix is given by

$$(g_{\mu\nu}) = \begin{bmatrix} (1 - r_g/r)^{-1} & 0 & 0 & 0 \\ 0 & r^2 & 0 & 0 \\ 0 & 0 & r^2 \sin^2\theta & 0 \\ 0 & 0 & 0 & -(1 - r_g/r)c^2 \end{bmatrix} , \quad (5.35)$$

and its inverse is

$$(g^{\mu\nu}) = \begin{bmatrix} e^\nu & 0 & 0 & 0 \\ 0 & 1/r^2 & 0 & 0 \\ 0 & 0 & 1/r^2 \sin^2\theta & 0 \\ 0 & 0 & 0 & -e^{-\nu}/c^2 \end{bmatrix} , \quad (5.36)$$

where we have defined

$$e^\nu = 1 - r_g/r . \quad (5.37)$$

Thus, the Hamilton-Jacobi for the motion of a free test particle moving in the pseudo-Riemann space defined by the metric (5.33) is given, using Eq. (5.36) in Eq. (5.32), as

$$e^\nu \left(\frac{\partial S}{\partial r}\right)^2 + \frac{1}{r^2} \left(\frac{\partial S}{\partial \theta}\right)^2 + \frac{1}{r^2 \sin^2\theta} \left(\frac{\partial S}{\partial \phi}\right)^2 - \frac{e^{-\nu}}{c^2} \left(\frac{\partial S}{\partial t}\right)^2 + \kappa^2 = 0 . \quad (5.38)$$

Defining p_1, p_2, p_3 and p_4 as

$$p_1 = \left(\frac{\partial S}{\partial r}\right) , \qquad p_2 = \left(\frac{\partial S}{\partial \theta}\right) , \qquad p_3 = \left(\frac{\partial S}{\partial \phi}\right) \qquad \text{and} \qquad p_4 = \left(\frac{\partial S}{\partial t}\right) ,$$

we have the function F (Eq. (4.63)) given by

$$F = e^\nu p_1^2 + \frac{p_2^2}{r^2} + \frac{p_3^2}{r^2 \sin^2 \theta} - \frac{e^{-\nu}}{c^2} p_4^2 + \kappa^2 = 0 \,. \tag{5.39}$$

The partial differentiations of this function are

$$F_r = \frac{r_g}{r^2} p_1^2 - 2\frac{p_2^2}{r^3} - 2\frac{p_3^2}{r^3 \sin^2 \theta} + \frac{e^{-2\nu}}{c^2} \frac{r_g}{r^2} p_4^2 \,,$$

$$F_\theta = -\frac{2p_3^2 \cos \theta}{r^2 \sin^3 \theta} \,,$$

$$F_\phi = 0 \,,$$

$$F_t = 0 \,,$$

$$F_S = 0 \,,$$

$$F_{p_1} = 2p_1 e^\nu \,,$$

$$F_{p_2} = 2p_2/r^2 \,,$$

$$F_{p_3} = 2p_3/r^2 \sin^2 \theta$$

and

$$F_{p_4} = -2e^{-\nu} p_4/c^2 \,.$$

Using these expressions and Eq. (4.71), we obtain the system of differential equations

$$\frac{dr}{2e^\nu p_1} = \frac{d\theta}{2p_2/r^2} = \frac{d\phi}{2p_3/r^2 \sin^2 \theta} = \frac{dt}{-2e^{-\nu} p_4/c^2}$$

$$= \frac{dS}{2e^\nu p_1^2 + 2p_2^2/r^2 + 2p_3^2/r^2 \sin^2 \theta - 2e^{-\nu} p_4^2/c^2}$$

$$= \frac{dp_1}{p_1^2 r_g/r^2 - 2p_2^2/r^3 - 2p_3^2/r^3 \sin^2 \theta + p_4^2 e^{-2\nu} r_g/c^2 r^2}$$

$$= \frac{dp_2}{-2p_3^2 \cos \theta/r^2 \sin^3 \theta} = \frac{dp_3}{0} = \frac{dp_4}{0} \,. \tag{5.40}$$

From the last two terms, we get for p_3 and p_4 the following

$$p_3 = a_1 \tag{5.41a}$$

and

$$p_4 = a_2 \,, \tag{5.41b}$$

where a_1 and a_2 are constants. Using Eq. (5.41a) in the terms containing dp_2 and integrating with the help of the second term of Eq. (5.40), we get

$$p_2^2 + \frac{a_1^2}{\sin^2 \theta} = a_3^2 \,, \tag{5.41c}$$

where a_3 is a constant. Now, substituting Eq. (5.41a), Eq. (5.41b) and Eq. (5.41c) in Eq. (5.39), we obtain

$$p_1^2 + \frac{e^{-\nu}}{r^2}a_3^2 - \frac{e^{-2\nu}}{c^2}a_2^2 + \kappa^2 e^{-\nu} = 0 \,. \tag{5.41d}$$

We can use Eq. (5.41) to integrate the differential 1-form in $\mathbb{R}^4$

$$dS = p_1 dr + p_2 d\theta + p_3 d\phi + p_4 dt \,,$$

obtaining the solution for Eq. (5.38) as

$$S = \pm \int \left[\frac{e^{-2\nu}}{c^2}a_2^2 - \frac{e^{-\nu}}{r^2}a_3^2 - \kappa^2 e^{-\nu} \right]^{1/2} dr \pm$$

$$\pm \int \left[a_3^2 - \frac{a_1^2}{\sin^2 \theta} \right]^{1/2} d\theta + a_1 \phi + a_2 t + b \,, \tag{5.42}$$

where a_3 and a_2 represent the angular momentum and the energy of the particle. The trajectory of the particle is obtained through the relation resulting from

$$\beta = \frac{\partial S}{\partial a_3} \,,$$

that is

$$\beta = \pm \int \frac{a_3 dr}{r^2 \left[\frac{a_2^2}{c^2} - \left(\frac{a_3^2}{r^2} + \kappa^2 \right) e^\nu \right]^{1/2}} \pm \int \frac{a_3 \sin \theta d\theta}{(a_3^2 \sin^2 \theta - a_1^2)^{1/2}} \,. \tag{5.43}$$

A physical discussion of the above solution (for the particular case $a_1 = 0$), can be found in the last above mentioned reference.

5.1.5 *Interaction of a Periodic Gravitational Wave with a Test Particle*

One of the solution of Einstein's equations in vacuum represents gravitational waves (see reference of I. Robinson and A. Trautman). The metric

$$ds^2 = dx^2 + dy^2 + 2(dt)(dr) - \sigma dt^2 \,, \tag{5.44}$$

represents one of such solutions. The representation here is given in the chart coordinate $\{x, y, r, t\}$, where r and t are related with the coordinate space z and time T as

$$r = (z - cT)/\sqrt{2} \tag{5.45a}$$

and

$$r = (z + cT)/\sqrt{2} \qquad (5.45b)$$

being c the speed of light ($c = 2.99792458 \times 10^8 meter/sec$). The function σ is given by

$$\sigma(x, y, t) = 2\omega(x^2 - y^2)\cos(\nu t), \qquad (5.46)$$

where ω represents the amplitude of the wave, and ν is the wave number. The matrix associated to this metric is

$$(g_{\mu\nu}) = \begin{pmatrix} 1 & 0 & 0 & 0 \\ 0 & 1 & 0 & 0 \\ 0 & 0 & 0 & 1 \\ 0 & 0 & 1 & -\sigma \end{pmatrix}, \qquad (5.47)$$

and its inverse matrix is

$$(g^{\mu\nu}) = \begin{pmatrix} 1 & 0 & 0 & 0 \\ 0 & 1 & 0 & 0 \\ 0 & 0 & \sigma & 1 \\ 0 & 0 & 1 & 0 \end{pmatrix}. \qquad (5.48)$$

Using Eq. (5.48) in Eq. (5.32), we obtain the Hamilton-Jacobi equation for a test particle under interaction with the gravitational wave,

$$\left(\frac{\partial S}{\partial x}\right)^2 + \left(\frac{\partial S}{\partial y}\right)^2 + 2\left(\frac{\partial S}{\partial r}\right)\left(\frac{\partial S}{\partial t}\right) + \sigma\left(\frac{\partial S}{\partial r}\right)^2 + \kappa^2 = 0. \qquad (5.49)$$

Defining p_1, p_2, p_3 and p_4 as

$$p_1 = \left(\frac{\partial S}{\partial x}\right), \quad p_2 = \left(\frac{\partial S}{\partial y}\right), \quad p_3 = \left(\frac{\partial S}{\partial r}\right), \quad \text{and} \quad p_4 = \left(\frac{\partial S}{\partial t}\right),$$

the function F is given by

$$F(x, y, r, t, S, \mathbf{p}) = p_1^2 + p_2^2 + 2p_3 p_4 + \sigma p_3^2 + \kappa^2 = 0. \qquad (5.50)$$

The partial differentiations of this function are

$$F_x = p_3^2 \sigma_x \,,$$
$$F_y = p_3^2 \sigma_y \,,$$
$$F_r = 0 \,,$$
$$F_t = p_3^2 \sigma_t \,,$$
$$F_S = 0 \,,$$
$$F_{p_1} = 2p_1 \,,$$
$$F_{p_2} = 2p_2 \,,$$
$$F_{p_3} = 2p_4 + 2\sigma p_3$$

and

$$F_{p_4} = 2p_3 \,.$$

Using these expressions in the system of equation (4.71), we have

$$\frac{dx}{2p_1} = \frac{dy}{2p_2} = \frac{dr}{2p_4 + 2\sigma p_3} = \frac{dt}{2p_3} = \frac{dS}{2p_1^2 + 2p_2^2 + 4p_3 p_4 + 2\sigma p_3^2}$$
$$= \frac{dp_1}{-p_3^2 \sigma_x} = \frac{dp_2}{-p_3^2 \sigma_y} = \frac{dp_3}{0} = \frac{dp_4}{-p_3^2 \sigma_t} \,. \tag{5.51}$$

From the term containing dp_3, it follows

$$p_3 = a = \text{constant} \,. \tag{5.52}$$

Using this result in the terms of Eq. (5.51) containing dt, dp_1 and dp_2, we obtain the following equations

$$\frac{dx}{dt} = \frac{1}{a} p_1 \,, \tag{5.53a}$$

$$\frac{dp_1}{dt} = -\frac{a}{2} \sigma_x \,, \tag{5.53b}$$

$$\frac{dy}{dt} = \frac{1}{a} p_2 \tag{5.54a}$$

and

$$\frac{dp_2}{dt} = -\frac{a}{2} \sigma_y \,. \tag{5.54b}$$

Doing the differentiation of Eq. (5.53a) and Eq. (5.54a) with respect to the variable t and using Eq. (5.53b), Eq. (5.54b), and the definition (5.46), we obtain an equivalent system given by

$$\frac{d^2x}{dt^2} + 2\omega x \cos(\nu t) = 0 \tag{5.55}$$

and

$$\frac{d^2y}{dt^2} - 2\omega y \cos(\nu t) = 0. \tag{5.56}$$

From the terms of Eq. (5.51) having dS and dt, we obtain the equation

$$\frac{dS}{dt} = \frac{1}{a}[p_1^2 + p_2^2 + 2p_3p_4 + \sigma p_3^2],$$

but using Eq. (5.50) and Eq. (5.52) here, we get

$$\frac{dS}{dt} = -\kappa^2/a. \tag{5.57}$$

From the terms of Eq. (5.51) having dr and dt, we have

$$dr = \frac{1}{a}(p_4 + \sigma p_3)dt,$$

where we can use Eq. (5.50), Eq. (5.52) to integrate and to bring about the expression

$$r = -\frac{1}{2a^2} \int [p_1^2 + p_2^2 + \sigma a^2 + \kappa^2]dt + \int \sigma dt + r_0. \tag{5.58}$$

The solution of Eq. (5.55) and Eq. (5.56) gives the evolution variables x and y as a function of the parameter t. Then, from Eq. (5.46), Eq. (5.53a) and Eq. (5.54a), the evolution of σ, p_1 and p_2 as a function of the parameter t can be known. This allows us to integrate Eq. (5.58) to know the evolution of r as a function of the parameter t. Finally, the evolution of the variable z and T can be known using Eq. (5.45a) and Eq. (5.45b),

$$z = (t + r)/\sqrt{2} \tag{5.59a}$$

and

$$cT = (t - r)/\sqrt{2}. \tag{5.59b}$$

From Eq. (5.59b), we can calculate the term dT/dt, and since the components of the velocity of the particle are given by

$$v_i = \frac{dx_i}{dT} = \frac{dx_i}{dt}\left(\frac{dT}{dt}\right)^{-1}, \tag{5.60}$$

we are able to completely know the behavior of the particle under interaction with this gravitational wave. A physical discussion about this solution can be found in reference G. López.

5.2 Trajectory of a Ray of Light

The electromagnetic field is governed by Maxwell's equations. When these equations are solved for vacuum, some of the solutions are plane waves which are characterized by the property that their direction of propagation and amplitude are the same everywhere

$$f = ae^{i(\mathbf{k}\cdot\mathbf{r}-\omega t+\alpha)},\tag{5.61}$$

where a is the amplitude, $\mathbf{k}$ is the wave vector pointing toward the direction of propagation, ω is the angular frequency of oscillation, α is a constant phase, and f represents any component of the electromagnetic field (the real part of Eq. (5.61) has this physical meaning). This type of wave are the basis for the subject geometric optics. Wherever the wave is not plane, but geometric optics is still applicable, we can write

$$f = ae^{i\psi},\tag{5.62}$$

where ψ is called the Eikonal and satisfies the equation

$$\eta^{\mu\nu}\left(\frac{\partial\psi}{\partial x^{\mu}}\right)\left(\frac{\partial\psi}{\partial x^{\nu}}\right) = 0,\tag{5.63}$$

where $(\eta^{\mu\nu})$ is the metric (5.29), and x^{μ} has the same meaning as above. In this way, Eq. (5.63) is the equation for the trajectory of a light ray in the flat space defined by the metric (5.30). Eq. (5.63) can be generalized to a pseudo-Riemann space through the metric (5.31). The equation for the trajectory of a light ray in a non-flat space becomes

$$g^{\mu\nu}(\mathbf{x})\left(\frac{\partial\psi}{\partial x^{\mu}}\right)\left(\frac{\partial\psi}{\partial x^{\nu}}\right) = 0.\tag{5.64}$$

As we can see from Eq. (5.28) and Eq. (5.32), Eq. (5.63) and Eq. (5.64) represent the Hamilton-Jacobi like equations for the particle with zero mass. In addition, the following approximation can be made if we still have a well defined frequency of oscillations at any time,

$$\psi(\mathbf{x}) = \Psi(x,y,z)e^{-i\omega(x,y,z)t}.\tag{5.65}$$

Assuming this approximation in Eq. (5.63), it follows that

$$\left(\frac{\partial\Psi}{\partial x}\right)^{2} + \left(\frac{\partial\Psi}{\partial y}\right)^{2} + \left(\frac{\partial\Psi}{\partial z}\right)^{2} = n^{2}(x,y,z),\tag{5.66}$$

where $n(x,y,z)$ is the refraction index, defined as

$$n(x,y,z) = \omega(x,y,z)/c.\tag{5.67}$$

The solution of Eq. (5.66) provides us with the geometric surface which makes up the front waves in a non-homogeneous medium characterized by the refraction index $n(x, y, z)$. Defining $p_i = \partial\Psi/\partial x_i$ for $i = x, y, z$ on this equation, this one can be written as

$$F = p_1^2 + p_2^2 + p_3^2 - n^2 = 0 \,,$$

which has the following equations for its characteristics

$$\frac{dx}{2p_1} = \frac{dy}{2p_2} = \frac{dz}{2p_z} = \frac{d\Psi}{2n^2} = \frac{dp_1}{2n_x n} = \frac{dp_2}{2n_y n} = \frac{dp_3}{2n_z n} \,,$$

where we have defined $n_i = \partial n/\partial x_i$. Note that if n does not depend on one particular variable, the associated p is automatically a constant (or a characteristic).

5.2.1 *Solution of the Eikonal Equation for a Refraction Index Depending on z*

Let us consider the case of stratified medium where the refraction index depends on the variable z only. Eq. (5.66) is written as

$$\left(\frac{\partial\Psi}{\partial x}\right)^2 + \left(\frac{\partial\Psi}{\partial y}\right)^2 + \left(\frac{\partial\Psi}{\partial z}\right)^2 = n^2(z) \,. \tag{5.68}$$

The solution of this equation can be easily found by proposing the function Ψ to be of the form

$$\Psi(x, y, z) = ax + by + u(z) \,, \tag{5.69}$$

where a and b are constant and the function $u(z)$ is unknown. Substituting Eq. (5.69) in Eq. (5.68) and after integration, it follows

$$u(z) = \int_0^z (n^2(\xi) - a^2 - b^2)^{1/2} d\xi \,. \tag{5.70}$$

So, the solution of Eq. (5.68) is

$$\Psi(x, y, z) = ax + by + \int_0^z (n^2(\xi) - a^2 - b^2)^{1/2} d\xi \,, \tag{5.71}$$

and the light rays in such a medium are found after taking the partial differentiations with respect the constant appearing in Eq. (5.71), that is using Eq. (1.78). These partial differentiations are

$$\frac{\partial\Psi}{\partial a} = x - a\int_0^z \frac{d\xi}{(n^2(\xi) - a^2 - b^2)^{1/2}} = \alpha$$

and

$$\frac{\partial \Psi}{\partial b} = y - b \int_0^z \frac{d\xi}{(n^2(\xi) - a^2 - b^2)^{1/2}} = \beta$$

or

$$x = \alpha + a \int_0^z \frac{d\xi}{(n^2(\xi) - a^2 - b^2)^{1/2}} = \alpha \tag{5.72a}$$

and

$$y = \beta + b \int_0^z \frac{d\xi}{(n^2(\xi) - a^2 - b^2)^{1/2}} = \beta. \tag{5.72b}$$

More discussion on this solution can be seen in reference K.R. Lunerburg and reference M. Born and E. Wolft.

5.2.2 *Solution of the Eikonal Equation for a Refraction Index Radially Depending*

In this case, we have a cylindrical symmetry, and Eq. (5.66) is given by

$$\left(\frac{\partial \Psi}{\partial \rho}\right)^2 + \frac{1}{\rho^2}\left(\frac{\partial \Psi}{\partial \theta}\right)^2 + \left(\frac{\partial \Psi}{\partial z}\right)^2 = n^2(\rho). \tag{5.73}$$

The easiest way to obtain the solution of this equation is proposing a solution of the form

$$\Psi(\rho, \theta, z) = a\theta + bz + R(\rho), \tag{5.74}$$

and substituting this in Eq. (5.73). The equation gotten for $R(\rho)$ is readily integrable, obtaining

$$R(\rho) = \pm \int_{\rho_0}^{\rho} \left(n^2(\xi) - \frac{a^2}{\xi^2} - b^2\right)^{1/2} d\xi, \tag{5.75}$$

and the solution for Eq. (5.73) is

$$\Psi(\rho, \theta, z) = a\theta + bz \pm \int_{\rho_0}^{\rho} \left(n^2(\xi) - \frac{a^2}{\xi^2} - b^2\right)^{1/2} d\xi. \tag{5.76}$$

The light rays in such a medium are given by (using $\partial \Psi/\partial a = \theta_0$ and $\partial \Psi/\partial b = z_0$)

$$\theta - \theta_0 = \pm a \int_{\rho_0}^{\rho} \frac{d\xi}{\xi^2 \left(n^2(\xi) - \frac{a^2}{\xi^2} - b^2\right)^{1/2}}, \tag{5.77a}$$

$$z - z_0 = \pm b \int_{\rho_0}^{\rho} \frac{d\xi}{\left(n^2(\xi) - \frac{a^2}{\xi^2} - b^2\right)^{1/2}}. \tag{5.77b}$$

A discussion of this solution can be found in the above last reference (for $b = 0$).

5.2.3   *Eikonal Equation for a Refraction Index Radially Depending in Sphere*

In this case, we have a spherical symmetry, and Eq. (5.66) is given by

$$\left(\frac{\partial\Psi}{\partial r}\right)^2 + \frac{1}{r^2}\left(\frac{\partial\Psi}{\partial\theta}\right)^2 + \frac{1}{r^2\sin^2\theta}\left(\frac{\partial\Psi}{\partial\phi}\right)^2 = n^2(r). \qquad (5.78)$$

Making the definitions $p_1 = \partial\Psi/\partial r$, $p_2 = \partial\Psi/\partial\theta$, and $p_3 = \partial\Psi/\partial\phi$, this equation is of the form

$$F = p_1^2 + \frac{p_2^2}{r^2} + \frac{p_3^2}{r^2\sin^2\theta} - n^2(r) = 0 , \qquad (5.79)$$

which has the following equations for its characteristics

$$\frac{dr}{2p_1} = \frac{d\theta}{2p_2/r^2} = \frac{d\phi}{2p_3/r^2\sin^2\theta} = \frac{d\Psi}{2n^2(r)} =$$

$$\frac{dp_1}{2p_2^2/r^3 + 2p_3^2/r^3\sin^2\theta - 2n_r n} = \frac{dp_2}{2\cos\theta p_3^3/\sin^3\theta} = \frac{dp_3}{0}$$

The last term implies that $p_3 = constant$ (or characteristic). Using this with the second and fifth terms and after integration, one gets the constant (or characteristic)

$$a = p_2^2 + \frac{p_3^2}{\sin^2\theta} . \qquad (5.80)$$

Using this in Eq. (5.79), one gets the relation

$$p_1^2 = n^2(r) - \frac{a}{r^2} . \qquad (5.81)$$

In this way, one can already integrate the equation $d\Psi = p_1 dr + p_2 d\theta + p_3 d\phi$ as

$$\Psi = \pm \int \sqrt{n^2(r) - \frac{a}{r^2}}\, dr \pm \int \sqrt{a - \frac{p_3^2}{\sin^2\theta}}\, d\theta + p_3\phi + b , \qquad (5.82)$$

where b is the constant of integration. Again, the partial differentiation of this expression with respect to the constants a and p_3 will bring about the trajectory of the ray.

5.3   Equivalent Hamiltonians

As it has been mentioned in Section 1.6.1, two Hamiltonian H and $\widetilde{H}$ are equivalents if there is a generatrix function F (canonical transformation) which is a non separable solution of a non linear partial differential equations (1.100) − (1.103). This particular application is interesting since it allows us to explore other method of solution for this equation which does not lead us to a separable form of the characteristics method. Let us consider the example of to find the generatrix function of the form $F_1(x, Q)$ which makes the Hamiltonian

$$H = \frac{p2}{2m} + \frac{1}{2}m\omega^2 x^2 \quad \text{and} \quad \widetilde{H} = \omega P \qquad (5.83)$$

equivalents. The function F_1 must be a solution of the non linear PDEFO (1.100) which is written for 1-D as

$$-\omega\frac{\partial F_1}{\partial Q} = \frac{1}{2m}\left(\frac{\partial F_1}{\partial x}\right)^2 + \frac{1}{2}m\omega^2 x^2 . \qquad (5.84)$$

Defining $\pi = (\partial F_1/\partial x)$ and $\theta = (\partial F_1/\partial Q)$, this non linear PDEFO is written as

$$F = \omega\theta + \frac{\pi^2}{2m} + \frac{1}{2}m\omega^2 x^2 = 0 . \qquad (5.85)$$

If one wants to used Clariut-Charpit method to find the solution of this equation, one would need to solve the linear equation for a function $U = U(x, Q, F_1, \pi, \theta)$ which would have the following equations for its characteristics

$$\frac{dx}{\pi/m} = \frac{dQ}{\omega} = \frac{dF_1}{\pi^2/m + \omega\theta} = \frac{d\pi}{-m\omega^2 x} = \frac{d\theta}{0} = \frac{dU}{0} , \qquad (5.86)$$

and because one must have that $\theta = c$ is one of the characteristic curves, $dF_1 = \pi dx + \theta dQ$ is of separable variable type, that is, it is of the form $F_1(x, Q) = f_1(x) + g_1(Q)$, which in turns means that $p = \partial F_1/\partial x = df_1(x)/dx$ and $P = -\partial F_1/\partial Q = dg_1(Q)/dQ$ which does not correspond to a canonical transformation between H and $\widetilde{H}$.

Therefore, let us assume that F_1 is of the form $F_1(x, Q) = f(x)g(Q)$ and do the substitution of this expression in Eq. (5.79). Then, Eq. (5.79) is written in the form

$$\omega f\frac{dg}{dQ} + \frac{g^2}{2m}\left(\frac{df}{dx}\right)^2 + \frac{1}{2}m\omega^2 x^2 = 0 . \qquad (5.87)$$

By selecting $f(x)$ of the form

$$f(x) = \lambda x^2 , \qquad (5.88)$$

one can make factorization

$$x^2\left\{\lambda\omega\left(\frac{dg}{dQ}\right) + \frac{\lambda^2}{2m}g^2 + \frac{1}{2}m\omega^2\right\} = 0 . \qquad (5.89)$$

So, for any $x \neq 0$, one gets the equation

$$\frac{dg}{dQ} = -\frac{\lambda}{2m\omega} g^2 - \frac{m\omega}{2\lambda} \tag{5.90}$$

which has the solution

$$g(Q) = -\frac{m\omega}{\lambda} \tan \frac{Q}{2} . \tag{5.91}$$

Thus, the solution obtained for Eq. (5.79) is given by

$$F_1(x, Q) = -m\omega x^2 \tan \frac{Q}{2} . \tag{5.92}$$

Let us see another example. Consider the parameter "s" (length of a circular accelerator) as the parameter of evolution of of the transversal motion of a bunch of charged particle in a circular accelerator. The equivalent Hamiltonians which describe this transversal motion are given by

$$H(x, p, s) = \frac{1}{2} \left(p^2 + K(s)x^2 \right) , \quad \text{and} \quad \tilde{H}(J, s) = \frac{J}{\beta(s)} , \tag{5.93}$$

where the function β satisfies the equation $2\beta\beta'' - \beta'^2 + 4K(s)\beta^2 = 4$ ($\beta' = d\beta/ds$), and $K(s)$ describes the magnetic elements (lattice) in the accelerator ring. Let us see this equivalence by looking a generatrix function of the form $F_1(x, Q, s)$, where $p = \partial F_1/\partial x$, $J = -\partial F_1/\partial Q$ and F_1 must satisfies the non linear PDEFO

$$-\frac{1}{\beta(s)} \frac{\partial F_1}{\partial Q} = \frac{1}{2} \left(\frac{\partial F_1}{\partial x} \right)^2 + \frac{1}{2} K(s)x^2 + \frac{\partial F_1}{\partial s} . \tag{5.94}$$

Assuming now F_1 of the form

$$F_1(x, Q, s) = \lambda x^2 \psi(Q, s) , \tag{5.95}$$

and substituting this expression into our non linear PDEFO, one gets the following quasi-linear PDEFO for ψ

$$\frac{1}{2} \frac{\partial \psi}{\partial Q} + \frac{\partial \psi}{\partial s} = -2\lambda\psi^2 - \frac{1}{2\lambda} K(s) . \tag{5.96}$$

The equations for its characteristics are given by

$$\beta(s)dQ = ds = \frac{d\psi}{-2\lambda\psi^2 - K(s)/2\lambda} . \tag{5.97}$$

From the first two term, one can get the characteristic

$$c_1 = Q - \int \frac{ds}{\beta(s)} , \tag{5.98}$$

and from the last two term, one has the following non linear ordinary differential equation

$$\frac{d\psi}{ds} + 2\lambda\psi^2 = -K(s)/2\lambda . \qquad (5.99)$$

Let us propose a solution for this equation of the form

$$\psi(Q,s) = \frac{1}{\beta(s)}\left[f(Q) + g(s)\right]. \qquad (5.100)$$

Then, since one has that $df/ds = (df/dQ)/\beta(s)$, because of the characteristic c_1 and after some arrangements, the substitution of Eq. (5.100) into Eq. (5.99) brings about the following equation for $f(Q)$,

$$\frac{df}{dQ} + 2\lambda f^2 = f(\beta' - 4\lambda g) - 2\lambda g^2 - \beta g' + g\beta' - \beta^2 K(s)/2\lambda . \qquad (5.101)$$

Choosing the function $g(s)$ of the form $g = \beta'/4\lambda$, this equation is written as

$$\frac{df}{dQ} + 2\lambda f^2 = \frac{1}{2\lambda}\left[\frac{\beta'^2}{4} - \frac{\beta\beta''}{2} - \beta^2 K(s)\right]. \qquad (5.102)$$

In this way, choosing now the β function such that it satisfies the equation

$$\frac{\beta'^2}{4} - \frac{\beta\beta''}{2} - \beta^2 K(s) = -1 . \qquad (5.103)$$

The equation for f is finally stablished as

$$\frac{df}{dQ} + 2\lambda f^2 + \frac{1}{2\lambda} = 0 . \qquad (5.104)$$

By selecting the λ value as $\lambda = 1/2$, the solution of this equation is just written as

$$f(Q) = -\tan Q . \qquad (5.105)$$

Thus, the generatrix function (5.95) is determined as

$$F_1(x,Q,s) = -\frac{x^2}{2\beta(s)}\left[\tan Q - \frac{\beta'(s)}{2}\right] , \qquad (5.106)$$

and the two Hamiltonian are equivalents.

5.4   Problems

5.1   Find the solution of the Hamilton-Jacobi equation of the one dimensional dissipative system of example 6 (Chap. 1, Sec. 1.6) which has the Hamiltonian

$$H = \frac{p^2}{2m} \exp(-2\alpha q/m),$$

and find the trajectory of the particle.

5.2   Show, in general, that the equation $L = vf(x,v) - H(x, f(x,v))$ is the envelope of the nonlinear partial differential equation

$$v\frac{\partial L}{\partial v} - L = H\left(x, \frac{\partial L}{\partial v}\right),$$

where $a = f(x,v)$ is the solution of

$$\frac{\partial H(x,a)}{\partial a} = v.$$

5.3   The magnetic field produced in the vacuum space between the two conductor of a coaxial cable can be characterized by the following expression for the potential $\mathbf{A}$ and ϕ

$$\mathbf{A} = \left(\frac{I}{2\pi}\theta, 0, 0\right)$$

and

$$\phi = 0,$$

where I is the current flowing in the conductors. Find the solution of the Hamilton-Jacobi equation for a charged particle moving in this field.

5.4   Solve Eq. (5.38) for the case $\nu = 0$.

5.5   Solve Eq. (5.49) for $\sigma = constant$.

5.6   Solve Eq. (5.66) in spherical coordinates for a refraction index depending on the radius.

5.7 Write the PDEFO defined in $\mathbb{R}^5$ for the generatrix function $F_1(\mathbf{x}, \mathbf{Q}, s)$ (see Eq. (1.100)) which establishes the equivalence between the Hamiltonians ("s" is the parameter of evolution of the system)

$$H = \frac{1}{2}\left(p_1^2 + K_1(s)x^2\right) + \frac{1}{2}\left(p_2^2 + K_2(s)y^2\right)$$

and

$$\tilde{H} = \frac{1}{\beta_1(s)}P_1 + \frac{1}{\beta_2(s)}P_2 \,.$$

ii) Propose a solution for this equation of the form

$$F = x^2 f_1(Q_1, s) + y^2 f_2(Q_2, s) \,,$$

and show (using the linearly independence of the variables x and y) that the functions f_1 and f_2 satisfy the quasi-linear PDEFO given by

$$\frac{1}{\beta_i}\frac{\partial f_i}{\partial Q_i} + \frac{\partial f_i}{\partial s} + K_i(s)/2 + 2f_i^2 = 0 \,, \quad i = 1, 2 \,.$$

iii) Show that the function

$$f_i(Q_i, s) = -\frac{1}{2\beta_i(s)}\left[\tan Q_i - \frac{\beta_i'(s)}{2}\right]$$

(where $\beta_i' = d\beta_i/ds$) is a solution of this quasi-linear PDEFO, and the function β_i satisfies the non linear ordinary differential equation

$$2\beta_i\beta_i'' - (\beta_i')^2 + 4K_i(s)\beta_i^2 = 4 \,, \quad i = 1, 2 \,.$$

5.8 A test particle is moving in the space-time manifold defined by the de Sitter metric

$$(g_{\mu\nu}) = f(x_\mu)\begin{pmatrix} 1 & 0 & 0 & 0 \\ 0 & 1 & 0 & 0 \\ 0 & 0 & 1 & 0 \\ 0 & 0 & 0 & -1 \end{pmatrix}\,, \quad f(x_\mu) = \left[1 - \lambda\left(x^2 + y^2 + z^2 - (ct)^2\right)^2\right]^{-2}\,.$$

Determine the Hamilton-Jacobi equation to find the trajectory of the particle in this space, and establish the equations for the characteristics.

5.5   *References*

5.1   H. Goldstein, *Classical Mechanics*, Addison-Wesley Press 1965. Chap. 9, pages 277-279 and 245-287.

5.2   L.C. Landau and E.M. Lifshitz, *The Classical Theory of Fields*, Pergamon Press, 1971. Cap. 7, pages 46-95.

5.3   C. Møller, *Theory of Relativity*, Oxford University Press, 1952 . Chap. IX.

5.4   I. Robinson and A. Trautman, Phys. Tev. Lett., **4**, No. 8, (1960) 431.

5.5   G. López, Internal Report IFUG-1988, Apd. Postal E-143, León Guanajuato, México.

5.6   R.K. Lunerburg, *Mathematical Theory of Optics*, University of california Press, 1964. Chap. I ,II.

5.7   M. Born and E. Wolf, *Principles of Optics*, Pergamon Press, 1980. Chap. III.

Chapter 6

Characteristic Surfaces of Linear Partial Differential Equation of Second Order

In this chapter we will use the PDEFO to find the so-called "characteristic surfaces of a linear partial differential equation of second order," and to see the usual classification of these equations. In the theory of Partial Differential Equation of Second Order (PDESO), the concept of "characteristic surfaces" has an important role in understanding the type of solution and its characteristics of existence.

6.1 Characteristic Surfaces of a Linear PDESO Defined in $\mathbb{R}^n$

For a linear PDESO defined in $\mathbb{R}^n$, one understands a expression of the form

$$\sum_{i,j=1}^{n} a_{ij}(x)\frac{\partial^2 u}{\partial x_i \partial x_j} = f(\mathbf{x}, u, \nabla u), \qquad (6.1)$$

where $\mathbf{x}$ is a point in $\Omega \subset \mathbb{R}^n, \mathbf{x} = (x_1, ..., x_n), a_{ij}$ is a once continuously differentiable function defined in $\Omega, i, j = 1, ..., n$. One will assume the symmetry $a_{ij} = a_{ji}$ for $i \neq j$. The function $u = u(\mathbf{x})$ will be at least twice continuously differentiable in Ω. The function f is defined $\mathbb{R}^{2n+1}$, and ∇u represents the expression $\nabla u = (u_1, ..., u_n)$ with $u_i = \partial u/\partial x_i$. One can make the following change of coordinates

$$\xi_i = \xi_i(\mathbf{x}) \quad i = 1, ..., n, \qquad (6.2)$$

where one needs to ask for the Jacobian of this transformation to be different from zero in Ω

$$\frac{\partial(\xi_1, ..., \xi_n)}{\partial(x_1, ..., x_n)} \neq 0 . \qquad (6.3)$$

With these variables, it follows that

$$\frac{\partial}{\partial x_i} = \sum_{k=1}^{n} \frac{\partial \xi_k}{\partial x_i} \frac{\partial}{\partial \xi_k} \qquad (6.4\text{a})$$

and

$$\frac{\partial^2}{\partial x_j \partial x_i} = \sum_{k=1}^{n} \frac{\partial^2 \xi_k}{\partial x_j \partial x_i} + \sum_{l,k=1}^{n} \frac{\partial \xi_k}{\partial x_i} \frac{\partial \xi_l}{\partial x_j} \frac{\partial^2}{\partial \xi_l \partial \xi_k} \,. \qquad (6.4\text{b})$$

Therefore, Eq. (6.1) is transformed to the expression

$$\sum_{i,j=1}^{n} a_{ij}(x) \sum_{k=1}^{n} \frac{\partial^2 \xi_k}{\partial x_j \partial x_i} + \sum_{i,j=1}^{n} a_{ij}(x) \sum_{l,k=1}^{n} \frac{\partial \xi_k}{\partial x_i} \frac{\partial \xi_l}{\partial x_j} \frac{\partial^2 \tilde{u}}{\partial \xi_l \partial \xi_k}$$

$$= f\left(\xi, \tilde{u}, \sum_{k=1}^{n} \frac{\partial \xi_k}{\partial x_1} \frac{\partial \tilde{u}}{\partial \xi_k}, ..., \sum_{k=1}^{n} \frac{\partial \xi_k}{\partial x_n} \frac{\partial \tilde{u}}{\partial \xi_k} \right), \qquad (6.5)$$

where $\tilde{u}$ is defined as

$$\tilde{u} = u(\mathbf{x}(\xi)).$$

Eq. (6.5) can be written as

$$\sum_{l,k=1}^{n} \alpha_{lk}(\xi) \frac{\partial^2 \tilde{u}}{\partial \xi_l \partial \xi_k} = g(\xi, \tilde{u}, \nabla \tilde{u}), \qquad (6.6)$$

where the functions α_{lk} and g have been defined as

$$\alpha_{lk} = \sum_{i,j=1}^{n} a_{ij}(x) \frac{\partial \xi_k}{\partial x_i} \frac{\partial \xi_l}{\partial x_j} \qquad (6.7\text{a})$$

and

$$g(\xi, \tilde{u}, \nabla_\xi \tilde{u}) = f\left(\xi, \tilde{u}, \sum_{k=1}^{n} \frac{\partial \xi_k}{\partial x_1} \frac{\partial \tilde{u}}{\partial \xi_k}, ..., \sum_{k=1}^{n} \frac{\partial \xi_k}{\partial x_n} \frac{\partial \tilde{u}}{\partial \xi_k} \right)$$

$$- \sum_{i,j=1}^{n} a_{ij}(\mathbf{x}) \sum_{k=1}^{n} \frac{\partial^2 \xi_k}{\partial x_j \partial x_i}. \qquad (6.7\text{b})$$

Now, due to (6.3) it is possible to choose the new variables such that all the elements on the diagonal of the matrix (α_{lk}) to be nulls, that is, $\alpha_{ll} = 0$ for $l = 1, ..., n$. From Eq. (6.7a), all of these elements satisfy the same nonlinear PDEFO,

$$\sum_{i,j=1}^{n} a_{ij}(\mathbf{x}) \frac{\partial \omega}{\partial x_i} \frac{\partial \omega}{\partial x_j} = 0, \qquad (6.8)$$

where $\omega = \omega(\mathbf{x})$. the solutions of this equation are called "characteristic surfaces" of (6.1). These n-surfaces, at the same time, can be used as the new set of variables (6.2). According to chapter 6 (section 2), Eq. (6.8) represents the following PDEFO

$$F(\mathbf{x}, \omega, \mathbf{p}) = \sum_{i,j=1}^{n} a_{ij}(\mathbf{x}) p_i p_j = 0, \tag{6.9}$$

which can be solved by the characteristics method shown in chapter 4. The equations for the characteristics of (6.9) are

$$\frac{dx_1}{2 \sum\limits_{i=1}^{n} a_{1i}(\mathbf{x}) p_i} = \ldots = \frac{dx_n}{2 \sum\limits_{i=1}^{n} a_{ni} p_i} = \frac{d\omega}{2 \sum\limits_{i,j=1}^{n} a_{ij}(\mathbf{x}) p_i p_j}$$

$$= \frac{dp_1}{-\sum\limits_{i,j=1}^{n} \left(\dfrac{\partial a_{ij}}{\partial x_1}\right) p_i p_j} = \ldots = \frac{dp_n}{-\sum\limits_{i,j=1}^{n} \left(\dfrac{\partial a_{ij}}{\partial x_n}\right) p_i p_j}. \tag{6.10}$$

Example 6.1. Let us find the characteristic surfaces of the equation

$$\sum_{i=1}^{3} \frac{\partial^2 u}{\partial x_i^2} - \frac{1}{c^2} \frac{\partial^2 u}{\partial t^2} = 0. \tag{6.11}$$

Defining the variable $x_0 = ct$, one has the following dependence $u = u(\mathbf{x})$ where $\mathbf{x} = (x_1, x_2, x_3, x_0)$. In this case one has

$$a_{ij}(\mathbf{x}) = 0, \quad \text{for} \quad i \neq j, \quad \text{and} \quad i, j = 0, 1, 2, 3. \tag{6.12a}$$

$$a_{ij}(\mathbf{x}) = 1, \quad \text{for} \quad i = 1, 2, 3 \quad \text{and} \quad a_{00}(\mathbf{x}) = -1. \tag{6.12b}$$

According to (6.7a) and (6.8), the equation for the characteristic surfaces is given by

$$\left(\frac{\partial \omega}{\partial x_1}\right)^2 + \left(\frac{\partial \omega}{\partial x_2}\right)^2 + \left(\frac{\partial \omega}{\partial x_3}\right)^2 - \left(\frac{\partial \omega}{\partial x_0}\right)^2 = 0. \tag{6.13}$$

This nonlinear PDEFO is written as

$$F(\mathbf{x}, \omega, \mathbf{p}) = p_1^2 + p_2^2 + p_3^2 - p_0^2 = 0, \tag{6.14}$$

and the equations for the characteristics are

$$\frac{dx_1}{2p_1} = \frac{dx_2}{2p_2} = \frac{dx_3}{2p_3} = \frac{dx_0}{-2p_0} = \frac{d\omega}{2(p_1^2 + p_2^2 + p_3^2 - p_0^2)}$$

$$= \frac{dp_1}{0} = \frac{dp_2}{0} = \frac{dp_3}{0} = \frac{dp_0}{0}.$$

These equations imply that

$$p_1 = b_1, \quad p_2 = b_2, \quad p_3 = b_3 \quad \text{and} \quad p_0 = b_0, \tag{6.15}$$

where b_j is a constant for $j = 0, 1, 2, 3, 4$. Eq. (6.14) gives the following relation among these constants

$$b_0 = \pm\sqrt{b_1^2 + b_2^2 + b_3^2}. \tag{6.16}$$

Since one has the relation

$$d\omega = p_1 dx_1 + p_2 dx_2 + p_3 dx_3 + p_0 dx_0, \tag{6.17}$$

one gets the solution

$$\omega(\mathbf{x}) = b_1 x_1 + b_2 x_2 + b_3 x_3 \pm \sqrt{b_1^2 + b_2^2 + b_3^2}\, x_0 + b_4. \tag{6.18}$$

Of course, the envelope of this family of surfaces is also a characteristic surface which can be calculated as shown in Chap. 5. Defining $\Phi = b_1 x_1 + b_2 x_2 + b_3 x_3 \pm \sqrt{b_1^2 + b_2^2 + b_3^2}\, x_0 + b_4 - \omega$ and doing $\Phi_{b_i} = 0$ for $i = 1, 2, 3$, one gets the so called "light cone" as enveloped for this equation (making $b_4 = 0$ and absorbing the constants in the definition of ω)

$$\omega(\mathbf{x}) = x_1^2 + x_2^2 + x_3^2 - x_0^2.$$

6.2 Characteristic Surfaces of a Linear PDESO Defined in $\mathbb{R}^2$

Let us see what happens for linear PDESO defined in some domain $\Omega \subset \mathbb{R}^2$. In this case, one gets for (6.1)

$$a_{11}(x, y)\frac{\partial^2}{\partial x^2} + 2a_{12}(x, y)\frac{\partial^2 u}{\partial x \partial y} + a_{22}(x, y)\frac{\partial^2 u}{\partial y^2} = f(x, y, u, u_x, u_y). \tag{6.19}$$

The new variables

$$\xi = \xi(x, y), \quad \eta = \eta(x, y) \tag{6.20}$$

bring about the changes

$$\frac{\partial}{\partial x} = \xi_x \frac{\partial}{\partial \xi} + \eta_x \frac{\partial}{\partial \eta}, \tag{6.21a}$$

$$\frac{\partial}{\partial y} = \xi_y \frac{\partial}{\partial \xi} + \eta_y \frac{\partial}{\partial \eta}, \tag{6.21b}$$

$$\frac{\partial^2}{\partial x^2} = \xi_{xx} \frac{\partial}{\partial \xi} + \eta_{xx} \frac{\partial}{\partial \eta} + \xi_x^2 \frac{\partial^2}{\partial \xi^2} + 2\xi_x \eta_x \frac{\partial^2}{\partial \xi \partial \eta} + \eta_{xx} \frac{\partial^2}{\partial \eta^2}, \tag{6.22a}$$

$$\frac{\partial^2}{\partial y^2} = \xi_{yy} \frac{\partial}{\partial \xi} + \eta_{yy} \frac{\partial}{\partial \eta} + \xi_y^2 \frac{\partial^2}{\partial \xi^2} + 2\xi_y \eta_y \frac{\partial^2}{\partial \xi \partial \eta} + \eta_{yy} \frac{\partial^2}{\partial \eta^2} \tag{6.22b}$$

and

$$\frac{\partial^2}{\partial x \partial y} = \xi_{xy} \frac{\partial}{\partial \xi} + \eta_{yx} \frac{\partial}{\partial \eta} + \xi_x \xi_y \frac{\partial^2}{\partial \xi^2} + (\xi_x \eta_y + \xi_y \eta_x) \frac{\partial^2}{\partial \xi \partial \eta} + \eta_x \eta_y \frac{\partial^2}{\partial \eta^2}. \tag{6.22c}$$

Eq. (6.19) can be written in terms of ξ and η as

$$\alpha_1(\xi, \eta) \frac{\partial^2 \tilde{u}}{\partial \xi^2} + 2\beta(\xi, \eta) \frac{\partial^2 \tilde{u}}{\partial \eta \partial \xi} + \alpha_2(\xi, \eta) \frac{\partial^2 \tilde{u}}{\partial \eta^2} = g(\xi, \eta, \tilde{u}, \tilde{u}_\xi, \tilde{u}_\eta), \tag{6.23}$$

where the functions α_1, β, α_2, and g are defined as

$$\alpha_1 = a_{11} \xi_x^2 + 2a_{12} \xi_x \xi_y + a_{22} \xi_{yy} \tag{6.24a}$$

$$\alpha_2 = a_{11} \eta_x^2 + 2a_{12} \eta_x \eta_y + a_{22} \eta_{yy} \tag{6.24b}$$

$$\beta = a_{11} \xi_x \eta_x + a_{12} (\xi_x \eta_y + \xi_y \eta_x) + a_{22} \xi_y \eta_y \tag{6.24c}$$

and

$$g = \tilde{f}(x(\xi, \eta), y(\xi, \eta), \tilde{u}, \xi_x \tilde{u}_\xi + \eta_x \tilde{u}_\eta, \xi_y \tilde{u}_\xi + \eta_y \tilde{u}_\eta) - a_{11}(\xi_{xx} \tilde{u}_\xi + \eta_{xx} \tilde{u}_\eta)$$
$$- 2a_{12}(\xi_{xy} \tilde{u}_\xi + \eta_{xy} \tilde{u}_\eta) a_{22}(\xi_{yy} \tilde{u}_\xi + \eta_{yy} \tilde{u}_\eta). \tag{6.24d}$$

One can select $\alpha_1 = \alpha_2 = 0$, which brings about, for the functions ξ and η, the following nonlinear PDEFO

$$a_{11}(x, y) \left(\frac{\partial \omega}{\partial x}\right)^2 + 2a_{12}(x, y) \left(\frac{\partial \omega}{\partial x}\right) \left(\frac{\partial \omega}{\partial y}\right) + a_{22}(x, y) \left(\frac{\partial \omega}{\partial y}\right)^2 = 0. \tag{6.25}$$

This equation cab be seen as a quadratic algebraic equation for the term ω_x (or for ω_y) whose roots are

$$\omega_x^{(\pm)} = -\frac{a_{12} \pm \sqrt{a_{12}^2 - a_{11} a_{22}}}{a_{11}} \omega_y. \tag{6.26}$$

This expression defines a linear PDEFO given by

$$\frac{\partial \omega}{\partial x} + \lambda^{(\pm)}(x, y)\frac{\partial \omega}{\partial y} = 0, \tag{6.27a}$$

where the function $\lambda^{(\pm)}$ is defined as

$$\lambda^{(\pm)}(x, y) = \frac{a_{12} \pm \sqrt{a_{12}^2 - a_{11}a_{22}}}{a_{11}}. \tag{6.27b}$$

From (6.23), the resulting linear PDESO is called **standard expression** of the linear PDESO (6.19) and is given by

$$2\beta(\xi, \eta)\frac{\partial^2 \tilde{u}}{\partial \eta \partial \xi} = g(\xi, \eta, \tilde{u}, \tilde{u}_\xi, \tilde{u}_\eta). \tag{6.28}$$

Eqs. (6.27) are used to classify PDESO (6.19) in the following way:

a) if $a_{12}^2 > a_{11}a_{22}$, the roots λ^+ and λ^- are reals and different, therefore there are two real characteristics curves. This is called hyperbolic case and one says that Eq. (6.19) is **hyperbolic**.

b) if $a_{12}^2 = a_{11}a_{22}$, the roots λ^+ and λ^- have the same expression (a_{12}/a_{11}) which is real, therefor there is only one characteristic curve. This is called parabolic case and one says that Eq. (6.19) is **parabolic**.

c) if $a_{12}^2 < a_{11}a_{22}$, the roots λ^+ and λ^- are complex and $\lambda^+ = (\lambda^-)^*$, therefor, there is not real characteristics. This is called elliptic case and one says that Eq. (6.19) is **elliptic**.

Example 6.2. Let us find the characteristic surfaces of the equation

$$\frac{\partial^2 u}{\partial x^2} - x\frac{\partial^2 u}{\partial y^2} = 0. \tag{6.29}$$

In this case, one has the following functions

$$a_{11} = 1, a_{12} = 0, a_{22} = -x.$$

From, Eq. (6.27b), it follows that

$$\lambda^+ = \sqrt{x}, \lambda^- = -\sqrt{x},$$

The partial differential equation for the characteristics surfaces is (Eq. (6.27a))

$$\frac{\partial \omega}{\partial x} \pm \sqrt{x}\frac{\partial \omega}{\partial y} = 0.$$

The equations for the characteristics curves are

$$dx = \frac{dy}{\pm\sqrt{x}} = \frac{dw}{0}.$$

From these equations one gets two characteristics curves which, at the same time, are the characteristics surfaces of (6.29). These characteristic surfaces are

$$\xi = y - \frac{2}{3}x^{3/2} \quad \text{and} \quad \eta = y + \frac{2}{3}x^{3/2},$$

and they form the new coordinates for the transformation which leads to the standard form of Eq. (6.29). The Jacobian of the transformation is given by

$$\frac{\partial(\xi,\eta)}{\partial(x,y)} = 2\sqrt{x}.$$

Therefor, for $x \neq 0$, one gets Eq. (6.29) standard form

$$\frac{\partial^2 \tilde{u}}{\partial\xi\partial\eta} = \frac{\tilde{u}_\eta - \tilde{u}_\xi}{3(\eta - \xi)}.$$

Note that for $x > 0$ Eq. (6.29) is of hyperbolic type; for $x = 0$ Eq. (6.29) is of parabolic type; and for $x < 0$ Eq. (6.29) is of elliptical type.

6.3   *Problems*

6.1   find the characteristic surfaces of the second order linear equation

$$\sum_{i=1}^{3} \frac{\partial^2 u}{\partial x_i^2} - \frac{1}{D}\frac{\partial u}{\partial t} = 0,$$

where D is constant.

6.2   Find the characteristic surfaces of the equation

$$x\frac{\partial^2 u}{\partial x^2} - \frac{\partial^2 u}{\partial y^2} = 0.$$

6.3   Find the characteristic surface and their standard form of the equations

i) $\dfrac{\partial^2 u}{\partial x^2} - \dfrac{\partial^2 u}{\partial y^2} = 0$.

ii) $\dfrac{\partial^2 u}{\partial x^2} + \dfrac{\partial^2 u}{\partial y^2} = 0$.

iii) $\dfrac{\partial^2 u}{\partial x^2} - \dfrac{\partial u}{\partial y} = 0$.

6.4   Determine the regions of the real plane $\mathbb{R}^2$ where the following PDESO

$$x\frac{\partial^2 u}{\partial x^2} - y\frac{\partial^2 u}{\partial y^2} = 0$$

is hyperbolic, elliptic, or parabolic, and find the characteristic surfaces and the standard form for the hyperbolic case.

6.4   References

6.1   F.H. Miller, *Partial Differential Equations*, John Wiley & Sons, 1941. Chap. II.

6.2   R. Courant and D. Hilbert, *Methods of Mathematical Physics, Vol. II*, John Wiley & Sons, 1962. Chap. I, II

6.3   V.S. Vladimirov, *Equations of Mathematical Physics*, Marcel Dekker, Inc. N.Y., 1971.

Index